THE DESIGNER'S GUIDE TO VERILOG-AMS

THE DESIGNER'S GUIDE BOOK SERIES

Consulting Editor
Kenneth S. Kundert

Books in the series:

The Designer's Guide to Verilog-AMS
ISBN: 1-4020-8044-1

The Designer's Guide to SPICE AND Spectre*
ISBN: 0-7923-9571-9

THE DESIGNER'S GUIDE TO VERILOG-AMS

First Edition

June 2004

KENNETH S. KUNDERT
Cadence Design Systems

OLAF ZINKE
Cadence Design Systems

Springer Science+Business Media, LLC

Library of Congress Cataloging-in-Publication

Title: The Designer's Guide to Verilog-AMS
Author (s): Kenneth S. Kundert & Olaf Zinke
ISBN 978-1-4757-8159-5 ISBN 978-1-4020-8045-6 (eBook)
DOI 10.1007/978-1-4020-8045-6

Originally published by Kluwer Academic Publishers in 2004
Softcover reprint of the hardcover 1st edition 2004

Printed on acid-free paper.

Contents

Preface

The Verilog Hardware Description Language (Verilog-HDL) has long been the most popular language for describing complex digital hardware. It started life as a proprietary language but was donated by Cadence Design Systems to the design community to serve as the basis of an open standard. That standard was formalized in 1995 by the IEEE in standard 1364-1995. About that same time a group named Analog Verilog International formed with the intent of proposing extensions to Verilog to support analog and mixed-signal simulation. The first fruits of the labor of that group became available in 1996 when the language definition of Verilog-A was released. Verilog-A was not intended to work directly with Verilog-HDL. Rather it was a language with similar syntax and related semantics that was intended to model analog systems and be compatible with SPICE-class circuit simulation engines. The first implementation of Verilog-A soon followed: a version from Cadence that ran on their Spectre circuit simulator.

As more implementations of Verilog-A became available, the group defining the analog and mixed-signal extensions to Verilog continued their work, releasing the definition of Verilog-AMS in 2000. Verilog-AMS combines both Verilog-HDL and Verilog-A, and adds additional mixed-signal constructs, providing a hardware description language suitable for analog, digital, and mixed-signal systems. Again, Cadence was first to release an implementation of this new language, in a product named AMS Designer that combines their Verilog and Spectre simulation engines. At the time this preface was written, all but the oldest commercial circuit simulators support Verilog-A, and each of the major ICCAD vendors offer mixed-signal simulators that support Verilog-AMS. Verilog-A is extensively used in both device modeling for circuit simulation and for behavioral modeling of analog systems and adoption of Verilog-AMS is growing rapidly.

Verilog-AMS is continuing to evolve. Version 2.1 of the Verilog-AMS standard is based on the IEEE Verilog 1364-1995 standard. It was released in January 2003. The committee charged with the development of Verilog-AMS (*www.eda.org/verilog-ams*) is currently working to improve and update the standard. Progress is currently being made to update the basis of the standard to the latest version of Verilog-HDL, IEEE 1364-2001. They are also working to integrate Verilog-AMS into SystemVerilog. Finally, extensions are being added to support compact semiconductor models and table models.

The intent of Verilog-AMS is to let designers of analog and mixed-signal systems and circuits create and use models that describe their designs. Once a design is described in Verilog-AMS, simulators are used to help designers better understand and verify their designs. Verilog-AMS allows designs to be described at the same level as does SPICE, but at the same time allows designs to also be described at higher more abstract levels. This range is needed for the larger more complex mixed-signal designs that are becoming commonplace today.

This book starts in Chapter 1 with a brief introduction to hardware description languages in general and Verilog-AMS in particular. Chapter 2 presents a formal top-down design methodology. While not used extensively today, top-down design is widely believed to be the only methodology available that can efficiently handle large complex mixed-signal designs. This chapter presents a refined and proven top-down methodology that overcomes many of the problems with existing top-down methodologies. Chapter 3 and Chapter 4 introduce the Verilog-A and Verilog-AMS languages. The important concepts of the languages are presented using practical and easy to understand examples. These chapters are intended to be read from beginning to end and are designed to take engineers with a working knowledge of programming concepts to the point where they are comfortable writing a wide range of Verilog-A and Verilog-AMS models. However, they do not cover all the details of the languages. Chapter 5 is a reference guide to the languages. It presents all of the details, but not in a completely linear fashion. Though it can be read from beginning to end, it was written with the expectation that most would use it as a reference, looking up just the details they need when they need them. As such, it, as with the rest of the book, is extensively cross referenced and indexed.

A word about the conventions used in this book. As new ideas and definitions are presented, a few keywords will be set in ***bold italics*** to make them easier to find and to call your attention to them as important points. Code is set in a sans serif font with keywords in **bold** and comments in *italics*. When in text, identifier names are set in *italics*. Acronyms that are spoken as words rather than letters are set in small caps; for example, SPICE. Besides the normal cross references found in the text, you will also find references that appear like this: (5§2.3p157). These abbreviated references include the chapter number, the section number, and finally the page number. Finally, all models presented in this book have been verified with the simulators from Cadence, either Spectre or AMS Designer as appropriate.

This book has two companion websites on which you can find updated information about both this book and its subject matter. *www.designers-guide.com* contains information about the book, including an errata sheet, the latest versions of the models given in this book, articles that contain additional information about both modeling and Verilog-AMS, and links to other sites that would be of interest. In addition, it also provides a discussion forum where you can ask questions and have conversations with other practicing design engineers. *www.verilog-ams.com* provides a burgeoning library of high quality user contributed Verilog-A and Verilog-AMS models.

It is our intention to continually update and improve this book. As such, we would like to ask for your help in the process. Please send your comments, suggestions, experiences, feedback and reports of errors to either *ken@designers-guide.com* or describe them at *www.designers-guide.com/Forum*.

Ken Kundert
Olaf Zinke
April 1, 2004

1 Introduction

1 Hardware Description Languages

Hardware description languages (HDLs) exist to describe hardware. In this they differ from traditional programming languages, which generally exist to describe algorithms. Programming languages such as C grew up with computers that were constructed with a Von Neumann architecture, meaning that they had a central processing unit (CPU) connected to memory that held both instructions and data. The CPU also controlled peripheral elements such as displays, keyboards, long-term storage, networking, etc. The fact that there was one CPU meant that programming languages developed to describe procedures that consist of a sequence of operations performed in a serial manner to the data in memory or on the peripheral elements. Contrast this with typical hardware systems where there are many individual components that all operate simultaneously. To properly describe hardware, one must be able to describe both the behavior of the individual components as well as how they are interconnected.

Hardware description languages have two primary applications: simulation and synthesis. With simulation, one applies various stimuli to an executable model that is described using the HDL in order to predict how it will respond. Simulation allows you to understand how complex systems behave before you incur the time and expense of implementing them. Synthesis is the process of actually implementing the hardware. Here the assumption is that the HDL is used to describe the hardware at an abstract level using component models that do not yet have a physical implementation, and that synthesis is the act of creating a new refined description with equivalent behavior at the inputs and outputs that uses components that do have a physical implementation. The goal for HDLs used for simulation is expressiveness: they should be able to describe a wide variety of behaviors easily. The goal for HDLs used for synthesis is realizability: they should only allow those behaviors that can be converted into an implementation to be described. As such, if a single language is used for both simulation and synthesis, then generally synthesis only supports a relatively constrained subset of the language.

Currently only digital finite-state machines are automatically synthesized. In this case, the desired behavior is described at the register-transfer level (RTL) using a well-defined subset of an HDL. Synthesis then converts the RTL description to an optimized gate-level description. Implementations of the gates are available from a library of standard cells.

Automated synthesis of analog or mixed-signal systems from a description of its desired behavior has not progressed to the point where it is practical except in a few very restricted cases. Furthermore, it is not clear that it will ever reach this point. For this reason, automated synthesis is not discussed in this book. Rather, the focus is on manual synthesis, the process undertaken by designers to convert high-level design requirements to an implementation that meets those requirements. This process, also known as the design process, is not one that traditionally uses hardware description languages when it involves the design of analog or mixed-signal systems. However, as mixed-signal systems become more complex there comes a time where it becomes impractical to design them without using abstraction. It is this point where use of HDLs becomes necessary as they are used to express the abstraction.

There are currently two HDLs available for describing mixed-signal hardware: Verilog-AMS and VHDL-AMS. As the names imply, they are extensions to the traditional Verilog and VHDL digital HDLs that are intended to support modeling of analog and mixed-signal systems. Though these languages have different strengths and weaknesses, they are intended to be used on the same types of circuits, in the same ways, to produce the same results. As such, they are competitors. To a large degree, the choice between them is currently determined by what language is being used for the digital part of the system. However, in the future the simulators supporting the HDLs are expected to fully support both languages, allowing the various components of a single system to be described with which either language is convenient. At that point, one language may begin to dominate over the other. In the mean time, both languages need supporting material that teaches designers how to use them. This book is intended to fulfill that role for Verilog-AMS, with other books doing the same for VHDL-AMS [15, 24].

2 The Verilog Family of Languages

Verilog-AMS is a modeling language for mixed-signal systems. It is primarily designed to support simulation of mixed-signal systems by allowing the system to be described to the simulator. However, mixed-signal systems represents a very broad class of systems and must support a wide variety of situations. As such, Verilog-AMS is a language that has a diverse range of capabilities.

The term "mixed-signal" suggests systems made up of parts that process digital signals and parts that process analog signals. As such, Verilog-AMS is a language that supports the description of both digital and analog components. Verilog-AMS is the merger and extension of two languages, Verilog-HDL and Verilog-A. These three languages currently make up the Verilog® family of languages.† Verilog-HDL allows the description of digital components and Verilog-A allows the description of analog. Verilog-AMS combines these two languages and adds additional capability to allow the description of mixed-signal components. The term Verilog-AMS will be used when referring to just the full AMS extensions and Verilog-A/MS when referring to both Verilog-A and Verilog-AMS.

Digital signals are discrete-event signals with discrete values. In other words, they are signals that are constant for a period of time, and then abruptly change to a new value. With digital signals there are generally only a small number of possible signal values, typically two, designated *true* and *false*, *high* and *low*, or *zero* and *one*. The Verilog-HDL language was designed to handle such signals, and the systems that generate them. This language has been available for many years. It is both well known and well documented, and so will not be discussed in depth in this book. If you wish more information on Verilog-HDL, try picking up one of the many books available that focus on it exclusively [1, 5, 23, 27].

Analog signals are signals that vary continuously, meaning that the value of the signal at any point may be any value from within a continuous range of values. There are two ways in which this typically occurs, as shown in Figure 1. Either the signal is piecewise constant versus time, meaning that it holds its value for a period of time before jumping to a new value, or it is continuous versus time, meaning that its value varies smoothly as a function of time. The former signals are referred to as being analog discrete-event signals and the latter are continuous-time signals. The figure shows the analog discrete-event signal jumping between values at regular intervals, but this is not necessary. Both the value, and the time at which the jump-events occur can be irregular.

Verilog-A is designed to allow modeling of systems that process continuous-time signals. While it can also handle systems that process the other types of signals, it is not efficient for doing so. Verilog-A has been around for many years, though not nearly as many as Verilog-HDL. The documentation available is either incomplete [8] or hard to find [28], and so Verilog-A is presented in depth in Chapter 3.

Since Verilog-AMS combines Verilog-HDL and Verilog-A, as shown in Figure 2, it inherits the ability to handle systems that process both digital and continuous-time

† Verilog is a registered trademark of Cadence Design Systems licensed to Accellera.

FIGURE 1 *Digital, analog discrete-event and analog continuous-time signals.*

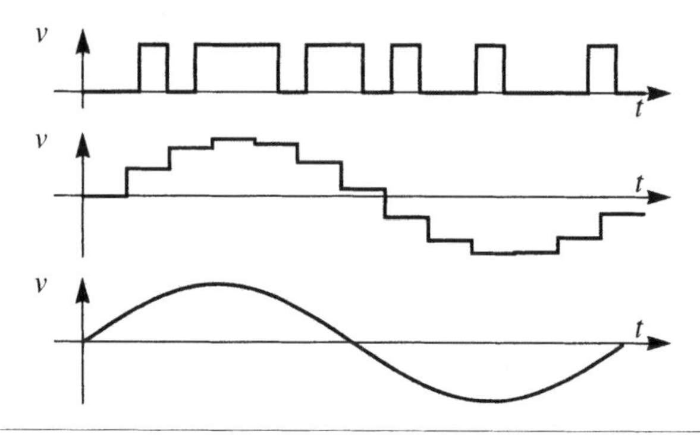

analog signals. It also adds the ability to efficiently support systems that process analog discrete-event signals. Verilog-AMS is the subject of Chapter 4,

FIGURE 2 *The relationship between Verilog-AMS, Verilog-A and Verilog-HDL.*

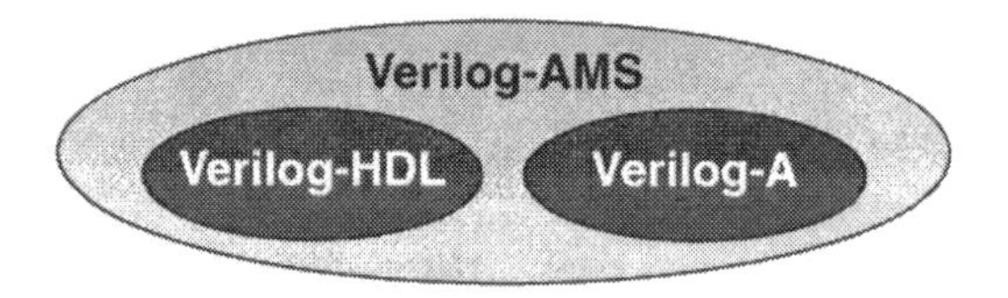

Verilog-AMS is expected to have a big impact on the design of mixed-signal systems because it provides a single language and a single simulator that is shared between analog and digital designers, and between block designers and system designers. It will be much easier to provide a single design flow that naturally supports analog, digital and mixed-signal blocks, making it simpler for these designers to work together.

Verilog-AMS makes it substantially more straight-forward to write behavioral models for mixed-signal blocks, and brings strong event-driven capabilities to analog simulation, allowing analog event-driven models to be written that perform with the speed and capacity inherited from the digital engines. This is very important, because most of the analog and mixed-signal models used in high-level simulations are naturally written using event-driven constructs. For example, blocks like ADCs, DACs, PLLs, $\Sigma\Delta$ converters, discrete-time filters (switched-capacitor), etc. are easily and very efficiently modeled using the analog event-driven features of the AMS languages.

Finally, it is important to recognize that Verilog-AMS is primarily used for verification. Unlike the digital languages, the AMS languages will not be used for synthesis in the foreseeable future because the only synthesis that is available for analog circuits is very narrowly focused.

3 Mixed-Signal Simulators

Mixed-signal simulators, by their very nature, combine two different methods of simulation: event-driven simulation as found in logic simulators, and continuous-time simulation as found in circuit simulators. As such, they are said to have two kernels; a discrete-event kernel and a continuous-time kernel. These two kernels are an essential feature of any mixed-signal simulator. Indeed, it is what separates mixed-signal simulation from other types of simulation. Within these constraints, mixed-signal simulators have changed considerably through the years.

Mixed-signal simulators first established themselves in the early 1990's. At this time there were two basic approaches, as shown in Figure 3. In one, a mixed-signal kernel was added to an established circuit simulator. Analogy's Saber and Georgia Tech's XSPICE are examples. These simulators offered relatively simple and low-level mechanisms to support event-driven simulation. They were quite different from, and incompatible with, the standard logic simulators of the day, such as Verilog-XL. As a result, while useful, these capabilities never gained wide acceptance. This lack of acceptance was addressed in the other approach, which simply glued together an established circuit simulator, generally some form of SPICE, and an established logic simulator, usually Verilog-XL. An example of this type of simulator is Spectre/Verilog-XL from Cadence. This approach addressed the lack of acceptance issue, but created ease-of-use and performance problems. The ease-of-use problems stem from the complexity of getting two simulators with very different use models to operate together. Generally, some form of mixed-signal design environment is required to manage the process of splitting the netlist between SPICE and Verilog, inserting the interface components, setting up and running the simulation, and accessing and displaying the results. Even with the environment, glued simulators developed a reputation of being difficult to use. The performance issues stem from the distant separation between the analog and digital parts of the circuit, and the overhead in the communication between the two simulators.

The AMS languages, Verilog-AMS and VHDL-AMS, address many of these issues. They provide a single standard input language that both supports mixed-signal descriptions, and is based on the standard languages used by logic simulators. As such, they provide the advantages of both types of early mixed-signal simulators,

FIGURE 3 *Architectures of early mixed-signal simulators. The environment, which is required with glued approach, is not shown.*

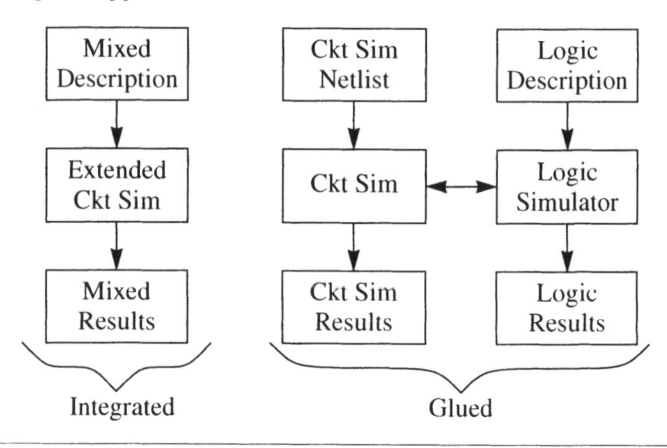

FIGURE 4 *Architectures of AMS mixed-signal simulators. The environment, which is required with glued approach, is not shown.*

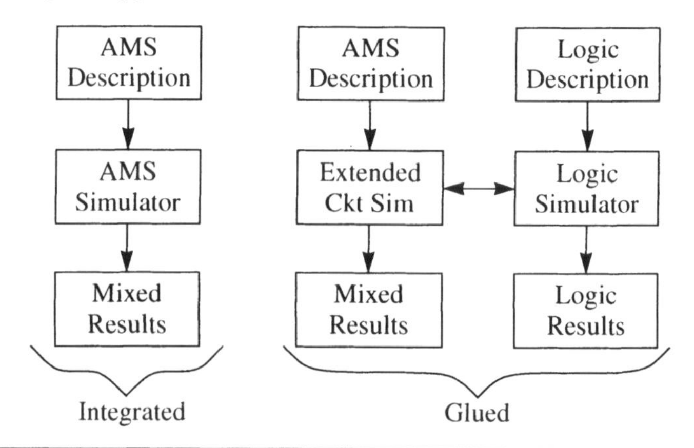

without the disadvantages. However, even with the AMS simulators, there is variety in the way they are constructed, as shown in Figure 4.

In an interesting twist, the integrated approach merges two established simulators whereas the glued approach adds an event-driven kernel to an established circuit simulator. The reason for this reversal stems from the need for both compatibility and performance when simulating modern mixed-signal systems-on-chip (MS-SOC) designs. The reason compatibility is critical is that generally large MS-SOCs are assembled from pieces that are designed with the help of the component simulators. The digital blocks would have been simulated with an established logic simulator and the analog blocks with an established circuit simulator. The various descriptions of the blocks must be reusable in the mixed-signal simulator without modification. Hence, the analog kernel must be completely compatible with the established circuit simulator, and the digital kernel must be completely compatible with the established logic simulator. In addition, because of the size of the designs, both kernels must provide performance that is comparable with the best engines on the market.

Both requirements point to the need to bolt together an established circuit simulator with an established logic simulator, and all commercial offerings do just that. However, it takes a great deal of effort to tightly merge an established circuit simulator and an established logic simulator, which is what is required when providing an integrated AMS simulator. Both the circuit and the logic simulator must be extensively modified, which requires a substantial commitment from both simulator development teams. Cadence's AMS Designer is an example of an integrated AMS simulator; it tightly merges the Spectre circuit simulator with the NC-SIM logic simulator.

The heavy investment needed to build an integrated AMS simulator has led to the development of glued AMS simulators. Here, AMS extensions are added to an established circuit simulator to provide the needed discrete-event kernel with tight integration to the continuous-time kernel, but extensions provide neither the performance of an established logic simulator, nor compatibility with such a simulator. As such, an established logic simulator must also be added to the mix, but in this case it is added as a separate, loosely integrated executable. In fact, the integration between the AMS simulator and the established logic simulator is virtually the same as the integration in the old glued mixed-signal simulators; with many of the same issues and with the added complication of having a redundant logic simulation capability.

Mentor's ADVance MS is an example of a glued AMS simulator. In another interesting twist, Mentor's marketing literature has described ADVance MS as a "single kernel" simulator. This is clearly incorrect, at least when using the word "kernel" in the way it is commonly used in mixed-signal simulation. An AMS simulator needs at least two kernels.

In an application of the old adage: "if you can't fix it, feature it," they are trying to suggest that by building the AMS extensions into their circuit simulator they have an inherently tighter coupling between the continuous-time and discrete-event kernels. But, in fact, there is no justification for such a claim. They are also trying to take attention off the fact that they have only a distant integration with their established ModelSim logic simulator. In a very real sense, they have a three kernel rather than a single kernel simulator, and two of the kernels are largely redundant. In the long run, expect Mentor to abandon this architecture and move to a more integrated architecture.

4 Applications of Verilog-AMS

There are five main reasons why engineers use Verilog-AMS:

1. to model components,
2. to create test benches,
3. to accelerate simulation,
4. to verify mixed-signal systems, and
5. to support the top-down design process.

It is important to understand each, as each represents an important application of Verilog-A/MS. Each will be briefly discussed with top-down design covered in detail in Chapter 2.

4.1 Component Modeling

A traditional circuit simulator such as SPICE provides a limited set of built-in models, those needed to model the components commonly available on integrated circuits, and provide relatively limited ability to add new models. They generally offer end users only the ability to add components described by a small number of simple formulas. These models are quite limited. They are also interpreted and so execute relatively slowly. As a result, this way of adding models is not suitable for complicated models or those that are heavily used.

In contrast, Verilog-A/MS provides a very wide variety of features and can be used to efficiently describe a broad range of models. Examples include …

1. Basic components such as resistors, capacitors, inductors, etc.
2. Semiconductor components such as diodes, BJTs, MOSFETs, varactors, etc. Public domain Verilog-A models exist for virtually all of the commonly used compact models such as the Gummel-Poon, VBIC, and Mextram BJT models and the MOS3, BSIM3 & 4, and EKV MOS models [29].
3. Functional blocks such as data converters, de/modulators, samplers, filters, etc.
4. Multi-disciplinary components such as sensors, actuators, transducers, etc.
5. Logic components and blocks such as gates, latches, registers, etc.
6. Test bench components such as sources and monitors.

Examples of all of these can be found in the library of Verilog-A/MS models at *www.verilog-ams.com*.

The ability to add models to a circuit simulator such as SPICE dramatically increases it range, and makes it immensely more powerful. The most obvious benefit is that it can simulate a broader variety of systems. The ability to add models of laser and photo diodes makes it suitable for electro-optical systems. The ability to add magnetic and power semiconductor models makes it suitable for power systems. The list is endless. Mechanical, thermal, acoustic, fluidic, etc. models are all possible. These capabilities are being used to model Micro-Electro-Mechanical Systems (MEMS), where multi-disciplinary components are constructed on the same silicon die. Even new electrical models, such as varactors, PIN diodes, etc. can be added to allow simulation of a wider variety of purely electrical systems.

In addition, as discussed in the remainder of this chapter, Verilog-A/MS can be used to add models of high-level blocks, allowing larger systems to be simulated; it allows models of digital components to be added, allowing simulation of mixed-signal systems; and it allows modeling of complicated test benches; allowing for more sophisticated tests.

4.1.1 Compact Models in Verilog-A

The term 'compact model' refers to the lumped models used by circuit simulators. More specifically, it is generally used to refer to the models of semiconductor components commonly incorporated into SPICE. As of the year this book was written, 2004, these models are largely implemented in C code that is compiled into the simulator. These models can be quite complex and are difficult to create and maintain. In addition, as each simulator has its own compiled model interface, the models are not portable between simulators. These issues create numerous problems for the people that develop the compact models, the people that develop the simulators, and the people that use and support the simulators [17]. As a result, there is increasing interest in instead writing and distributing the models in Verilog-A. Doing so addresses most, if not all, of the existing problems. Verilog-A is a standard language, so models would be portable. If the models were made available in source form, end users could correct any flaws in the model or enhance them as needed. Finally, models written in Verilog-A tend to be much smaller and more easily maintained than models written in C. Usually more than a 10× reduction in the amount of source code needed to represent the model is easily achievable.

To support the development of compact models in Verilog-A, extensions to the language are currently being developed. It is hoped that a future edition of this book will describe those extensions.

4.2 Test Benches

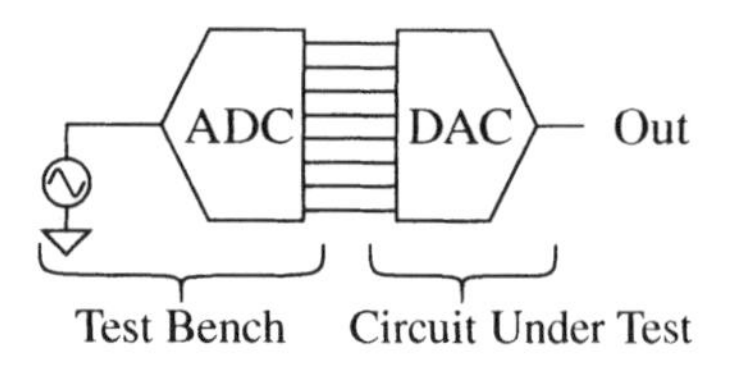

The term 'test bench' is used to refer to the circuitry that is added to the circuit under test so as to provide it with an environment in which it can properly operate. The test bench acts as a replacement for the circuitry that the circuit under test will be connected to when it is in normal operation. At a minimum it consists of one or more sources, but could be quite complicated. For example, as shown in the figure, when testing a digital-to-analog converter (DAC), one might include an ideal analog-to-digital converter (ADC) in the test bench as a way of generating the digital patterns used to drive the DAC. Conversely, when testing an ADC, one might use an ideal DAC to convert the output back to an analog signal that can be easily compared to the input, or further processed. Such represents a simple example of what could be done. In many cases it might be desirable for the test bench to be a high level model of the system in which the circuit under test is expected to operate. This is common when the interaction between the circuit and its environment is complex and employs sophisticated feedback; as might be the case when the circuit under test is adjustable and its test bench is expected to perform a calibration step before going on to measure the circuit's performance.

In these cases, it is natural to describe either the entire test bench, or the components that make up the test bench, with Verilog-A/MS. This is often relatively easy to do because the test bench is constructed with idealized components, so no time need be spent on trying to model non-idealities. This is certainly true with the test bench for the DAC, where the ADC used would be perfectly linear, and so no time is expended trying to match the nonlinearity present in a real ADC.

4.3 Simulation Acceleration

With designs becoming larger and their behavior becoming more complex, it is taking longer and longer to verify them with simulation. This is particularly true if the design is described at the transistor level. Often, there are particular critical portions of the design that are of the most concern. In this case, the simulation time can be reduced if the non-critical portions of the circuit are replaced with behavioral models. A behavioral model for a block will simulate much faster than a transistor-level model of the same block. If enough of the circuit is converted to behavioral-level, the resulting simulations can be much faster (2§4.3p27). For example, a complex PRML read channel required more than a month to complete simulations when represented completely at the transistor level, and over night when all blocks except the one of primary concern were represented by behavioral models. In the extreme, the idea of using Verilog-A/

MS to accelerate simulation is closely related to the idea of using Verilog-A/MS to model test benches, as the portion of the circuit represented by behavioral models effectively acts as a test bench for the portion represented at the transistor level.

4.4 Mixed-Signal Design

When designing mixed-signal circuits Verilog-AMS is very useful as it allows both digital and analog circuits to be described in a way that is most suitable for each type of circuit. With digital circuits, either gate- or behavioral-level Verilog-HDL is used, and with analog circuits, either transistor- or behavioral-level Verilog-A/MS is used. The fact that Verilog-AMS is a superset of Verilog-HDL means that the representations for the two parts of the circuit can be easily combined. But it also means that both analog and digital designers can operate with languages that are familiar and comfortable for them, and yet they can still work together.

4.5 Top-Down Design

Top-down design is a design methodology that is useful when designing large complex systems. The basic premise is to design and verify the system at an abstract or 'block diagram' level before starting the detailed design of the individual blocks. Top-down design would be used in lieu of the more traditional approach referred to as *bottom-up design*.

In bottom-up design, one fully designs the individual blocks before focusing on the design of the block diagram of the system. Bottom-up design generally requires that the individual blocks be over designed so that when connected together to form the system there is enough margin to overcome unexpected problems. The big risk being that the required system performance might not be achievable with the blocks as designed, meaning that one or more blocks would have to be re-designed.

With top-down design, the individual block performance needed to meet the overall system performance requirements is carefully studied and understood before the blocks are developed. This reduces the need for over design in the individual blocks, but at the risk that the anticipated performance for one or more blocks is unachievable, which would require that the system design be revisited.

Clearly, to reduce the expense and time required for rework, there must be extensive communication between the system and block level designers. This is where Verilog-AMS becomes useful. Verilog-AMS enables a much richer form of communication; one where models can be exchanged in addition to words, diagrams, and specifications. It is often said that a picture is worth a thousand words. In the same way, a model is even more powerful than a picture because it is a working example of what is expected. With a model, one can run scenarios to clarify meaning and intent. Furthermore, representative models can be used as proxies for the implemented blocks, meaning that the system designers can try out their designs with models from the block designers, or block designers can try out their designs in a model of the larger system. In either case, communication is improved and the frequency and severity of errors is reduced along with the need for rework.

What's Next

The first four uses of Verilog-A/MS are relatively straight-forward extensions of existing design methodologies, and as such, can be employed relatively easily. In some sense, the main challenge is learning to use the language. The language is presented in Chapters 3-5. However, the last use, top-down design, represents a fundamental change in the way that design is done. As such, its adoption will be much more difficult and slow. The next chapter discusses the need for top-down design and how it might be done.

2 Top-Down Design

1 Mixed-Signal Design Productivity

The mixed-signal design process has changed relatively little over the past two decades, and in comparison to the digital design process, is slow, labor intensive, and error prone. While digital designers have improved their design methodology and adopted design automation, analog and mixed-signal designers by and large have not.

There are two reasons why digital designers are far ahead of analog designers in improving their design processes. First, digital designers confronted the need to design very large and complex systems much earlier than analog designers. Consider that large digital chips today consist of tens of millions of transistors, while complex analog chips contain only tens of thousands of devices. Second, the digital design problem is much more amenable to automation than the analog problem.

Consider a digital design. In most cases digital systems are implemented as finite-state machines (FSM) and constructed from standard cell libraries. Using a FSM formulation acts to unify and homogenize digital design and gives it a well-understood mathematical foundation. This foundation was thoroughly explored in the late '80s resulting in the commercial logic synthesis tools of the early '90s. These tools take a register-transfer level description (RTL), a relatively high-level description of a digital system that is created by designers and can be verified with the help of logic simulators, to produce an optimized gate-level description of the system. This transformation is possible because digital systems are constructed from a limited set of relatively simple and well-behaved building blocks. The building blocks of digital systems are gates and registers. The blocks, generally referred to as cells, all share a common I/O model and so are easily interconnected, are derived from a relatively small number of cell types that have very simple and easily described behavior, are easily parameterized in terms of the number of inputs and outputs, and have a simple and easily adjusted performance trade-off that involves only speed and power. Logic synthesizers operate by creating a complete mathematical description upon which it performs transformations in order to create an optimal design in terms of speed, power, and area. This is a two step process. First, equivalence transformations are applied to the mathematical descriptions in order to reduce the total number of gates, which minimizes the area, and the depth of the logic, which roughly maximizes the speed. This is possible because each block has a simple logical description and a common interface model. Then, the drive ability of each gate is adjusted to provide the lowest power while still meeting speed requirements. This is possible because this speed-power trade-off is easily made in each gate.

Now consider analog design. Analog design has no equivalent to finite-state machines, and so has no unified formulation and no common mathematical foundation. It also has no universal equivalence transformations that allow the topology of a circuit to be easily modified without risk of breaking the circuit. These problems prevent a topological mapping from a behavioral description to hardware. Even if one were mathematically possible, the lack of a common I/O model for analog blocks would prevent the topological modifications that are needed for either mapping or topology optimization.

It might be possible to try to enforce a common I/O model for analog circuits, but doing so would be very expensive. For example, one might simply specify that the signals at the inputs and outputs of analog blocks center around a particular value, have the same maximum swing, and that outputs have zero output impedance and inputs have zero input admittance. The problem is that doing so would necessitate extra circuitry in each analog block that is there simply to assure compliance to the I/O model. That circuitry reduces the overall performance of the circuit by increasing power dissipation, increasing noise, decreasing bandwidth, etc. This differs from the digital world where the common I/O model was achieved naturally and without significant cost. In addition, it is not possible to achieve these ideals at high frequencies. Instead, some common reference impedance would have to be specified, such as the 50Ω used at the system level, but driving such loads greatly increases power dissipation.

Finally, there is no simple way to trade-off the various performance metrics that are important with analog blocks, which makes it difficult to perform a parametric optimization. Sensitivity-based local optimizations can be used, but the improvement provided by these approaches is usually small. Monte Carlo-based global optimizers offer better improvements, but require substantially more in the way of computer resources.

The end result is that analog designers have no equivalent to RTL, a relatively high-level language in which they can describe their design and from which they can synthesize an implementation that is guaranteed to be functionally correct and have near optimal performance. As such they must transform their designs from concept to implementation by hand, and so the design process is naturally much slower and more error prone that the design process for digital circuits.

The outlook for providing the equivalent to logic synthesis for analog designers is bleak. However, things cannot continue as they are; the current situation is becoming untenable. While a complex digital chip can be designed correctly on the first try in a few months, designing a complex analog chip can require 3-4 re-spins and up to a year and a half to get right. This is problematic for many reasons:

1. The tremendous mismatch in schedule and risk between the analog and digital portions of a mixed-signal design makes it difficult to justify combining analog and digital on the same chip.
2. The high risk makes planning difficult. It is hard to predict when product will be available, and when valuable analog designers will free up.
3. A long time-to-market makes it tough to react to changes in market trends and competitive pressures.
4. Analog and mixed-signal product development demands large investments of time and money. This makes it difficult to justify developing new analog products, especially in tough economic times.
5. Analog and mixed-signal designers are scarce and hard to recruit. Compounding this problem is the inherently low-level of productivity of the current mixed-signal design process, which makes it difficult for small design houses that are not focused on analog to field an analog design capability.
6. Some mixed-signal designs are becoming so large that, with the low productivity of the analog design process, a team of analog designers that is large enough to take on the project and complete it in a timely manner simply cannot be assembled.
7. To compensate for semiconductor processes that are increasingly unfriendly to analog designs results in an increasing use of auto calibration and adaptive filtering. This substantially increases the complexity of both the design and the verification of the design, which magnifies the problems already mentioned.

Clearly a change is needed. It is interesting to note that when digital designers were trying to design systems of a size comparable to today's mixed-signal designs, their design process was not that different from what analog designers are using today. But it was at that point that they began to transition to a more structured and more automated design methodology. For analog designers, substantial automation may not be in the cards in the near future, but the need to transition to a more structured design methodology that is both more efficient and that allows designers to handle the growing size of analog and mixed-signal circuits is clearly needed.

The availability of logic synthesis tools was not the only enabling factor for digital designers to move to more efficient design methodologies. By moving to FSM and RTL, digital designers also gave up considerable performance in terms of speed and power. They made this sacrifice to be able to design the larger and more complex systems quickly. This sacrifice was a critically important enabling factor. Analog and mixed-signal designers have not demonstrated the willingness to make a similar sacrifice. In those cases where performance is not critical, the tendency is to instead convert the circuit to a digital implementation in order to gain flexibility. In the remaining cases sacrificing performance is not an option; however it is also not clear that such a sacrifice is needed. Analog designers do not have the equivalent of logic synthesis, so they will continue to use custom design methodologies. While moving to IP (intellectual property) reuse may entail some sacrifice in overall system performance, changing to a top-down design methodology does not inherently imply lower system performance. In fact, the opposite is usually the case, using top-down design results in higher performance. Rather, the sacrifice that is demanded of analog and mixed-signal designers is that they must learn new skills, such as behavioral modeling, and they must be more disciplined in the way they design.

It is unlikely that analog and mixed-signal designers will ever be allowed on a large scale to trade any substantial amount of performance and power for a reduction in design time. Rather, in those cases where the performance and power requirements are not demanding, a digital implementation is usually preferred.

2 Traditional Approaches to Mixed-Signal Design

At the Design Automation Conference in 1998, Ron Collett of Collett International presented findings from a 1997 productivity study in which his firm analyzed 21 chip designs from 14 leading semiconductor firms. The study revealed a productivity gap of 14× between the most and least productive design teams. The study also revealed that developing analog and mixed-signal circuitry requires three to seven times more effort per transistor than designing digital control logic, though this factor was normalized out of the 14× ratio.

The reason for the poor productivity of those at the bottom end of the scale are increasingly complex designs combined with a continued preference for a bottom-up design methodology and the occurrence of verification late in the design cycle, which leads to errors and re-spins. There's a huge disparity in productivity between those mixed-signal designers who have transitioned to an effective "top-down" design methodology, and those who practice "bottom-up" design and rely solely on SPICE.

2.1 Bottom-Up Design

The traditional approach to design is referred to as bottom-up design. In it, the design process starts with the design of the individual blocks, which are then combined to form the system. The design of the blocks starts with a set of specifications and ends with a transistor level implementation. Each block is verified as a stand-alone unit against specifications and not in the context of the overall system. Once verified individually, the blocks are then combined and verified together, but at this point the entire system is represented at the transistor level.

While the bottom-up design style continues to be effective for small designs, large designs expose several important problems in this approach.

1. Once the blocks are combined, simulation takes a long time and verification becomes difficult and perhaps impossible. The amount of verification must be reduced to meet time and compute constraints. Inadequate verification may cause projects to be delayed because of the need for extra silicon prototypes.
2. For complex designs, the greatest impact on the performance, cost and functionality is typically found at the architectural level. With a bottom-up design style, little if any architectural exploration is performed and so these types of improvements are often missed.
3. Any errors or problems found when assembling the system are expensive to fix because they involve redesign of the blocks.
4. Communication between designers is critical, yet an informal and error prone approach to communication is employed. In order to assure the whole design works properly when the blocks are combined, the designers must be in close proximity and must communicate often. With the limited ability to verify the system, any failure in communication could result in the need for additional silicon prototypes.
5. Several important and expensive steps in the bottom-up design process must be performed serially, which stretches the time required to complete the design. Examples include system-level verification and test development.

The number of designers that can be used effectively in a bottom-up design process is limited by the need for intensive communication between the designers and the inherently serial nature of several of the steps. The communication requirements also tend to require that designers be co-located.

2.2 Moving to Top-Down Design

In order to address these challenges, many design teams are either looking to, or else have already implemented, a top-down design methodology [4,18]. In a primitive top-down approach [3], the architecture of the chip is defined as a block diagram and simulated and optimized using a system simulator such as Matlab or Simulink. From the high-level simulation, requirements for the individual circuit blocks are derived. Circuits are then designed individually to meet these specifications. Finally, the entire chip is laid out and verified against the original requirements.

This represents the widely held view of what top-down design is. And while this is a step towards top-down design, it only addresses one of the issues with bottom-up design (point 2 in Section 2.1). In essence, these design groups have not fundamentally changed their design process; they have simply added an architectural exploration step to the front. The flaw in this approach is that there is an important discontinuity in the design flow that results because the representation used during the architectural exploration phase is incompatible with the representation used during implementation. This discontinuity creates two serious problems. First, it leaves the block designers without an efficient way of assuring that the blocks all work together as expected. One could assemble transistor-level representations of the blocks and run simulations, but the simulations are too slow to be effective. The first time the blocks can be thoroughly tested together is first silicon, and at that point any errors found trigger a re-spin. Second, the discontinuity makes communication more difficult and ad hoc and so acts to separate the system designers from the circuit designers, and the circuit designers from each other. Without a reliable communication channel, designers resort to using verbal or written specifications, which are often incomplete, poorly communicated, and forgotten half way through the project. It is the poor communication process that creates many of the errors that force re-spins, and the separation that allows the errors to hide until the design is available as silicon.

To overcome these issues, one needs a design methodology that

1. Improves communication between designers (between system and block designers, between block designers, and between current designers and future designers (to support reuse)).
2. Eliminates the discontinuity that acts to hide errors and separate the system designers from the block designers.
3. Improves verification so that it finds the errors that cause re-spins, and finds them earlier so that they are less disruptive and easier to fix.
4. Improves designer effectiveness.
5. Reorganizes the design tasks, making them more parallel and eliminating long serial dependencies.
6. Reduces the need for extensive transistor-level final verification.
7. Eliminates re-spins!

RF designers typically use this type of primitive top-down design approach. They begin with the system design. Typically using a spreadsheet, the gain, noise figure and distortion budget is explored; and with the help of guides like the Friis equation, is distributed amongst the various blocks of the receiver. The design is then iterated until the performance of the system as predicted by the spreadsheet is met and the performance requirements on the blocks are reasonable. At this point, the design proceeds bottom-up relying solely on transistor-level design. Eventually, the spreadsheet is updated with the actual values coming from the transistor-level simulation, and if the system performance is not satisfactory, the process repeats. The problem is that even when using the updated results, the performance predicted by the spreadsheet will not match the results achieved in silicon. This happens because of miscommunications, either in the meaning or the actual values of the block specification, and because the system-level description is crude and does not account for things like loading effects. When designing non-integrated receivers, this is not as problematic because all the stages are generally designed for power matching and the voltage supply is reasonably high ($V_{dd} \geq 5$ V). In CMOS design the voltage supply is low (1.2 V in a 0.13 μm process) and the blocks do not share matched impedances. The result, of course, is that multiple silicon iterations are needed to achieve the required system performance levels.

3 Principles of Top-Down Design

A well designed top-down design process methodically proceeds from architecture- to transistor-level design. Each level is fully designed before proceeding to the next and each level is fully leveraged in the design of the next. Doing so acts to partition the design into smaller, well defined blocks, and so allows more designers to work together productively. This tends to reduce the total time required to complete the design. A top-down design process also formalizes and improves communications between designers. This reduces the number of flaws that creep into a design because of miscommunication. The formal nature of the communication also allows designers to be located at different sites and still be effective.

Following a top-down design methodology also reduces the impact of changes that come late in the design cycle. If, for whatever reason, the circuit needs to be partially redesigned, the infrastructure put in place as part of the methodology allows the change to be made quickly. The models can be updated and the impact on the rest of the system can be quickly evaluated. The simulation plan and the infrastructure for mixed-level simulations is already available and can be quickly applied to verify any changes.

An effective top-down design process follows a set of basic principles.

1. A shared design representation is used for the entire length of the project that allows the design to be simulated by all members of the design team and in which all types of descriptions (behavioral, circuit, layout) can be co-simulated.
2. During the design process each change to the design is verified in the context of the entire previously verified design as dictated by the verification plan.
3. A design process that includes careful verification planning where risks are identified up-front and simulation and modeling plans are developed that act to mitigate the risks.
4. A design process that involves multiple passes, starting with high level abstractions and refining as the detail becomes available. In effect, running through the entire process very quickly at the beginning with rough estimates and guesses to get a better understanding and better estimates, and then refining the design as the process progresses.
5. To the degree possible, specifications and plans should be manifested as executable models and scripts, things that are used in the design process on a daily basis, rather than as written documents.

3.1 A Shared Design Representation

In the primitive top-down design process commonly used today, the system designers use a different design representation than the circuit designers. For example, the system designers might use a spreadsheet, Matlab, Simulink, SPW, or SystemView while the circuit designers would use Verilog, VHDL, or SPICE. This causes a myriad of problems, perhaps the most important being that they are using different tools to explore the design and that make it difficult for them to share what they learn during the design process. As mentioned before, this leads to communication problems and eventually to design errors that are generally not caught until first silicon.

If instead a common simulatable design representation is used, such as Verilog-AMS, then the system engineers can build an architectural-level description of the design constructed from behavioral models of each of the blocks that can be evaluated by each of the circuit designers. In effect, the circuit designers start by receiving an executable example of what they are expected to design. If they have trouble meeting their assigned specifications, they can go back to the system engineers with simulations that show how the system is affected by the shortfall. Since both types of engineers are working in a familiar environment, communication is enhanced and potential resolutions can be explored together. The ready availability of behavioral models of the blocks that act as executable examples greatly reduces the need for onerous specifications that describe the desired behavior of each block, specifications that are often poorly written and that frequently go unread.

3.2 Every Change is Verified

In a primitive top-down design methodology, the architectural description of the system is usually thoroughly verified using simulation. However, the design is then re-created at the circuit level during the implementation phase and this version of the design is never checked against the original architectural description. This discontinuity is where many of the errors creep in that are not found until first silicon. In effect, the verification that was done in the architectural phase is not leveraged during the implementation phase. Verification in the implementation phase is generally not as effective because it is slow and so cannot be as comprehensive. In addition, the test benches used during the architectural design phase often cannot be reused during the implementation phase, and are generally difficult to re-create.

It is important instead to use a common simulatable representation for the design that allows both the system-level and circuit-level descriptions of the various blocks to be co-simulated, as is possible with Verilog-AMS. This capability is referred to as mixed-level simulation [14,22]. With it, individual blocks or small sets of individual blocks can be represented at the transistor- or even layout-level and be co-simulated with the rest of the system, which is described with high-level behavioral models. While these simulations are often considerably slower than simulations where every block is described at the high-level, they are also considerably faster than simulations where every block is described at the transistor level. And they allow the individual blocks to be verified in a known-good representation of the entire system. In effect, the system simulations are leveraged to provide an extensively verified test bench for the individual blocks.

Consider a simple example. It is not uncommon for a system to fail at first silicon because of a miscommunication over the polarity of a digital signal, such as a clock, enable, or reset line. Such errors cannot survive in the high-level description of the system because of the extensive testing that occurs at this level. They also cannot survive during mixed-level simulation because the individual block, where the error is presumed to be, is co-simulated with shared models for which the polarity of the signal has already been verified. They can, however, survive in either a bottom-up or primitive top-down design process because the test benches for the individual blocks are created by the corresponding block designers. Any misunderstanding of the required interface for the block will be reflected both in the implementation of the block and in its test bench, and so will not be caught until first silicon.

3.3 Verification Planning

Generally users of bottom-up or primitive top-down design methodologies find that the errors detected at first silicon are a result of rather mundane mistakes that occur at the interfaces between the various blocks. These errors are generally caused by communication breakdowns and would have been easy to find with simulation had anyone thought to look for them. The fact is that the focus of verification efforts in these methodologies is on guaranteeing the performance of the individual blocks, and not on identifying the problems that result when the blocks are interconnected. Some effort is generally spent on trying to verify the system as a whole, but it comes late in the process when the system is represented largely at the transistor level. At this stage, the simulations are quite slow and the amount of functionality that can be verified is very limited.

In a well-conceived top-down design process a verification planning step occurs that focuses on anticipating and preventing the problems that occur when assembling the blocks into a system. In order to be effective, it must move the verification to as early in the design process as possible and occur with as much of the system described at a high level as possible. Moving the chip-level verification up in the design process means that errors are caught sooner, and so are easier and less expensive to fix. Using high-level models means that the simulations run faster, and so can be substantially more comprehensive.

In a zealousness to accelerate the simulation, care must be taken to assure that enough of the system is at the right level to assure that the desired verification is actually occurring. Thus, the verification plans must include both simulation plans, that describe how the verification is to occur, and modeling plans, that indicate which models need to be available to support the verification plan and which effects should be included in the models. The modeling plan is very important. Without it behavioral models may be written that do not include the desired effect while including many effects that are unrelated to what is being verified. If they do not model the desired effect, then the verification will not be effective, if they model too many effects, then the verification runs unnecessarily slow and the models become more difficult and expensive to develop. The goal with the modeling plan is to identify a collection of simple models along with directions as to when they should be used, rather that trying to develop one complex model that is used in all cases.

An important benefit of the verification plan is that it allows the design team to react to late changes in the design requirements with confidence. When a change to the requirements occurs, it is possible to quickly revisit the verification plan, modify the design, update the models, and then apply it to the amended design to assure it satisfies the new requirements. Since it spells out all the simulations that need to occur to verify the design, there is little chance that a change needed by the new requirements that happens to break some other part of the design will go unnoticed.

Another important benefit of the up-front planning process used when developing the verification plan is that it tends to sensitize the design team to possible problem areas, with the result that those areas are less likely to become problems.

3.4 Multiple Passes

To reduce the risk of design iterations that result from unanticipated problems, it is important to take steps to expose potential problems early by working completely through an abstract representation of the design, using estimates as needed. As the design progresses and more detailed and reliable information becomes available, the abstract representation is successively refined. This process begins by developing a top-level behavioral model of the system, which is refined until it is believed to be an accurate estimate of the desired architecture. At this point, there should be reasonable understanding as to how the blocks will be implemented, allowing size estimates to be made for the blocks, which leads to an initial floorplan. Top-level routing is then possible, which leads to parasitic estimation, with the effect of the parasitics being back annotated to the top-level. Simulations can then expose potential performance problems as a result of the layout, before the blocks are available. This may result in early changes to the architecture, changes in block specifications, or perhaps just an improvement of the verification plan. However, these changes occur early in the design process, which greatly reduces the amount of redesign needed.

As the blocks are implemented and more information becomes available, the process is continually repeated while updating and refining the design.

3.5 Executable Specifications and Plans

When a design fails because of miscommunications between engineers, it is a natural reaction to insist that in future designs, formal specifications and plans be written in advance as a way of avoiding such problems. In practice, this does not work as well as generally hoped. The act of writing things down is beneficial as it gets the engineers thinking more deeply about their designs up-front, and so they develop a better understanding of what is expected and what could go wrong. However, as for the written specifications and plans themselves, they can take a long time to write, are usually not very well written or maintained, and are often not read by the other engineers. The fact is, the documents themselves are rarely effective at improving the communications between the engineers. Rather it is the better understanding that comes from writing them that acts to improve communications.

If instead, specifications and plans took the form of executable models and scripts that would be used and valued on a daily basis, perhaps with a small amount of accompanying documentation, then they would be naturally well written, well used, and well maintained. The models and scripts are also inherently very specific, which eliminates the ambiguity that occurs in written documents and that can result in misunderstandings that lead to re-spins. These models and scripts should be maintained with the design data and shared between all designers. This avoids another of the problem with written specifications; the situation where one engineer is unaware that another has updated a specification.

Use of executable specifications and plans in the form of models and scripts both substantially improves the design process for the initial version of the chip, as well as greatly easing reuse of either the design as a whole, or the blocks used in constructing the chip. IP reuse, or reuse of the blocks, is made considerably easier because validated high-level models of the blocks are available at the end of the design process. These models would then be used to easily evaluate the blocks as to there suitability for use in other designs. Derivatives, or system reuse, are greatly simplified by the existence of all of the models and scripts. It makes it much easier for either a new team, or new team members, to get a quick understanding of an existing design and initiate the process of making changes to retarget the design to a new application. Furthermore, having models and verification scripts that have been refined by the experiences of the first design team make it more likely that the follow-on designs will debut without surprises.

4 A Rigorous Top-Down Design Process

The rigorous top-down design methodology described here is a substantial refinement of the primitive top-down process described in Section 2.2. It follows the principles described in Section 3 in order to address all of the problems associated with bottom-up design, as identified in Section 2.1.

4.1 Simulation and Modeling Plans

An important focus in a good top-down design methodology is the development of a comprehensive verification plan, which in turn leads to the simulation and modeling plans. The process begins by identifying particular areas of concern in the design. Plans are then developed for how each area of concern is to be verified. The plans specify how the tests are performed, and which blocks are at the transistor level during the test. For example, if an area of concern is the loading of one block on another, the plan might specify that one test should include both blocks represented at the transistor level together. For those blocks for which models are used, the effects required to be included in the model are identified for each test. This is the beginning of the modeling plan. Typically, many different models will be created for each block.

It is important to resist the temptation to specify and write models that are more complicated than necessary. Start with simple models and only model additional effects as needed (and as spelled out in the modeling plan). Also, the emphasis when writing models should be to model the behavior of the block, not its structure. A simple equation that relates the signals on the terminals is preferred to a more complicated model that tries to mimic the internal working of the block. This is counter to the inclination of most designers, whose intimate knowledge of the internal operation of the block usually causes them to write models that are faithful to the architecture of the block, but are more complicated than necessary.

It is also not necessary to model the behavior of a circuit block outside its normal operating range. Instead, you can add code to a model that looks for inappropriate situations and reports them. Consider a block that supports only a limited input bias range. It is not necessary to model the behavior of the block when the input bias is outside the desired range if in a properly designed circuit it will never operate in that region. It is preferable to simply generate a warning that an undesirable situation has occurred.

Following these general rules results in faster simulations and less time spent writing models. However, the question of how much detail is needed in each model is a delicate one that must be answered with great care. It is important to understand the imperfections in the blocks and how those imperfections affect the overall performance of the system before one can know whether the effects should be included in a model. Also, it is not always true that a pure behavioral model is superior to a more structurally accurate model. Often making the model more structurally accurate makes it more predictive, and also may make it easier to include some secondary effects due to parasitics.

The simulation plan is applied initially to the high-level description of the system, where it can be quickly debugged. Once validated, it can then be applied to transistor level simulations.

A formal planning process generally results in more efficient and more comprehensive verification, meaning that more flaws are caught early and so there are fewer design iterations.

4.2 System-Level Verification

System-level design is generally performed by system engineers. Their goal is to find an algorithm and architecture that implements the required functionality while providing adequate performance at minimum cost. They typically use system-level simulators, such as Simulink [21] or SPW [6], that allow them to explore various algorithms and evaluate trade-offs early in the design process. These tools are preferred because they represent the design as a block diagram, they run quickly, and they have large libraries of predefined blocks for common application areas.

This phase of the design provides a greater understanding of the system early in the design process [12,13]. It also allows a rapid optimization of the algorithm and moves trades to the front of the design process where changes are inexpensive and easy to make. Unworkable approaches are discarded early. Simulation is also moved further up in the design process where it is much faster and can also be used to help partition the system into blocks and budget their performance requirements.

Once the algorithm is chosen, it must be mapped to a particular architecture. Thus, it must be refined to the point where the blocks used at the system level accurately reflect the way the circuit is partitioned for implementation. The blocks must represent sections of the circuit that are to be designed and verified as a unit. Furthermore, the interfaces must be chosen carefully to avoid interaction between the blocks that are hard to predict and model, such as loading or coupling. The primary goal at this phase is the accurate modeling of the blocks and their interfaces. This contrasts with the goal during algorithm design, which is to quickly predict the output behavior of the entire circuit with little concern about matching the architectural structure of the chip as implemented. As such, Verilog-AMS becomes preferred during this phase of the design because it allows accurate modeling of the interfaces and supports mixed-level simulation.

The transition between algorithm and architecture design currently represents a discontinuity in the design flow. The tools used during algorithm design are different from the ones used during architecture design, and they generally operate off of different design representations. Thus, the design must be re-entered, which is a source of inefficiencies and errors. It also prevents the test benches and constraints used during the algorithm design phase from being used during the rest of the design process.

On the digital side, tools such as SPW do provide paths to implementation via Verilog and VHDL generation. Similar capabilities do not yet exist for the analog or mixed-signal portions of the design. An alternative is to use Verilog-AMS for both algorithm and architecture design. This has not been done to date because the simulators that support these languages are still relatively new. It will probably take a while for this approach to become established because of the absence of the application specific libraries needed for rapid system-level exploration. Alternatively, simulators like AMS Designer from Cadence that supports both algorithm and architecture development by combining SPW with Verilog-AMS can be used [2].

4.3 Mixed-Level Simulation

Without analog synthesis, analog design is done the old fashioned way, with designers manually converting specifications to circuits. While this allows for more creativity and gives higher performance, it also results in more errors, particularly those that stem from miscommunication. These miscommunications result in errors that prevent the system from operating properly when the blocks are assembled even though the blocks were thought to be correct when tested individually.

To overcome this problem, mixed-level simulation is employed in a top-down design methodology for analog and mixed-signal circuits. This represents a significant but essential departure from the digital design methodology. Mixed-level simulation is required to establish that the blocks function as designed in the overall system.

To verify a block with mixed-level simulation, the model of the block in the top-level schematic is replaced with the transistor level schematic of the block before running the simulation. For this reason, all of the blocks in the architectural description of the system must be "pin-accurate", meaning that they must have the right number of pins and characteristics of each pin must be representative of the expected signal levels, polarities, impedances, etc.

The pin-accurate system description, described at a high level, acts as a test bench for the block, which is described at the transistor level. Thus, the block is verified in the context of the system, and it is easy to see the effect of imperfections in the block on the performance of the system. Mixed-level simulation requires that both the system and the block designers use the same simulator and that it be well suited for both system- and transistor-level simulation.

Mixed-level simulation allows a natural sharing of information between the system and block designers. When the system-level model is passed to the block designer, the behavioral model of a block becomes an executable specification and the description of the system becomes an executable test bench for the block. When the transistor level design of the block is complete, it is easily included in the system-level simulation.

Mixed-level simulation is the only feasible approach currently available for verifying large complex mixed-signal systems. Some propose to use either timing simulators (sometimes referred to as fast or reduced accuracy circuit simulators) or circuit simulators running on parallel processors. However, both approaches defer system-level verification until the whole system is available at transistor level, and neither provides the performance nor the generality needed to thoroughly verify most mixed-signal systems. They do, however, have roles to play both within the mixed-level simulation process and during final verification.

Successful use of mixed-level simulation requires careful planning and forethought (provided during the verification planning process). And even then, there is no guarantee that it will find all the problems with a design. However, it will find many problems, and it will find them much earlier in the design process, before full-chip simulations, when they are much less costly to fix. And with mixed-level simulation, it is possible to run tests that are much too expensive to run with full-chip simulation.

4.3.1 Mixed-Level Simulation Example

Though this example is several years old, it is representative of the type of circuit complexity that is common today. It is a PRML channel chip that is difficult to simulate for two reasons. First, it is a relatively large circuit that involves both analog and digital sections that are closely coupled. Second, the architecture involves complex feedback loops and adaptive circuits that take many cycles to settle. The combination of many transistors and many cycles combines with the result being a simulation that is so expensive as to be impractical. In this case, the expected simulation time was predicted to be greater than a month.

The traditional approach to simulating a complex circuit like this is to simulate the blocks individually. Of course this verifies that the blocks work individually, but not together. In addition, for this circuit it is difficult to verify the blocks when operating outside the system and it is difficult to predict the performance of the system just knowing the performance of the individual blocks.

When the architecture was simulated at a high level with each block represented by a pin-accurate behavioral model, the simulation time was less than 10 minutes. Then, when a single block was run at the transistor level, the simulation ran overnight. Even though the full system was never simulated at the transistor level, when fabricated it worked the first time because this methodology verified the blocks in the context of the system and it verified the interfaces between the blocks.

4.4 Bottom-Up Verification

Once a block is implemented, one could update the models that represent it to more closely mimic its actual behavior. This improves the effectiveness of mixed-level and system-level simulation and is referred to as bottom-up verification. To reduce the chance of errors, it is best done during the mixed-level simulation procedure. In this way, the verification of a block by mixed-level simulation becomes a three step process. First the proposed block functionality is verified by including an idealized model of the block in system-level simulations. Then, the functionality of the block as implemented is verified by replacing the idealized model with the netlist of the block. This also allows the effect of the block's imperfections on the system performance to be observed. Finally, the netlist of the block is replaced by an extracted model. By comparing the results achieved from simulations that involved the netlist and extracted model, the functionality and accuracy of the extracted model can be verified. From then on, mixed-level simulations of other blocks are made more representative by using the extracted model of the block just verified rather than the idealized model.

Bottom-up verification should not be delayed until the end of the design process, but should rather be done continuously during the entire design process. Once a block has been implemented to the degree that a more representative model can be extracted, that model should replace the idealized top-level model as long as it does not evaluate substantially slower. Doing so tends to improve the effectiveness of mixed-level simulation and the accuracy of the extracted models. And, as a side benefit, the models that would be needed if the block were to be made into a shared IP block are already available and tested at the end of the project. If the model development for bottom-up verification were postponed to the end of the design process, the natural pressure to meet schedule targets as designs near tape-out often result in some of the verification, and perhaps all of the modeling, being skipped. This increases the chance of error and decreases the opportunity for reuse.

When done properly, bottom-up verification allows the detailed verification of very large systems. The behavioral simulation runs quickly because the details of the implementation are discarded while keeping the details of the behavior. Because the details of the implementation are discarded, the detailed behavioral models generated in a bottom-up verification process are useful for third-party IP evaluation and reuse.

4.5 Final Verification

In a top-down design process, SPICE-level simulation is used judiciously in order to get its benefits without incurring its costs. All blocks are simulated at the transistor level in the context of the system (mixed-level simulation) in order to verify their functionality and interfaces. Areas of special concern, such as critical paths, are identified up front in the verification plan and simulated at the transistor level. The performance of the circuit is verified by simulating just the signal path or key pieces of it at the transistor level. Finally, if start-up behavior is a concern, it is also simulated at the transistor level. The idea is not to eliminate SPICE simulation, but to reduce the time spent in SPICE simulation while increasing the effectiveness of simulation in general by careful planning.

It is in this phase that the dynamic timing simulators (fast reduced-accuracy transistor-level simulators) play an important role. They often have the capacity to simulate large mixed-signal systems at the transistor level for a reasonable period of time. Again, even with timing simulators the simulations are generally only fast enough to provide limited verification. So use of a timing simulator does not offset the need for mixed-level simulation.

4.6 Test

During the design phase, the test engineers should use top-level description of the design as a simulatable prototype upon which to develop the test plan and test programs. The availability of a working model of the system early in the design process allows test engineers to begin the development and testing of test programs early. Moving this activity, which used to occur exclusively after the design was complete, so that it starts at the same time the block design begins significantly reduces the time-to-production [9,10,26]. Bringing test development into the design phase can reduce post-silicon debug time by 50% and can eliminate a turn by finding chips that are untestable early. It can also improve tests, which then improves yield.

5 Further Benefits of Top-Down Design

Besides the benefits described in the previous two sections, a rigorous top-down design methodology addresses all of the various needs and issues described in Sections 1-2, which includes the following.

5.1 Improves Communications Between Engineers

Communications between the designers is improved in two substantial ways. First, the use of a shared high-level model of the system that everyone verifies their designs in eliminates most of the miscommunication that occurs when following either bottom-up or primitive top-down design processes. In addition, the executable specifications and plans (models and scripts) are more specific and less ambiguous, and considerably reduce the time spent writing and reading formal specifications, providing a more efficient and effective replacement.

5.2 Improves Productivity

The improved productivity that results with a rigorous top-down design process is due mainly to the elimination of mistakes and re-spins. A more formal, less error-prone design process with better communication between engineers means that less time is spent making and recovering from mistakes, and more time is spent on productive design tasks.

5.3 Improves Ability to Handle Complex Designs

The ability of a design team following a rigorous top-down design methodology to handle more complex designs follows from the better system exploration and from the increased understanding of the design that comes from it, and from the improved communications. In addition, the use of mixed-level simulation dramatically improves the team's ability to verify complex circuits.

5.4 Allows Parallel Execution of Design Tasks

Reducing time-to-market is an important way in which design teams can increase their chance of success and the returns of their products. Part of the reduction in time-to-market is a result of the improved productivity and effectiveness of the design team, as described above. However, a rigorous top-down design methodology also has the benefit in that it allows more engineers to be effectively engaged in the development process at the same time, resulting in a further decrease in time-to-market.

As pointed out earlier, the existence of a shared executable high-level model of the system allows the test program development to be done in parallel with the block design and assembly, thereby eliminating a large contributor to the delay between when the design team and when manufacturing think the chip is ready to go. In addition, many of the final verification tasks that are needed with a bottom-up design style are moved forward in the form of mixed-level simulations and performed in parallel by the block and top-level designers. The block developers can also get started developing models or evaluating IP while the system designer is finalizing the overall system architecture.

The improved and more formal communication that results in a rigorous top-down design methodology allows more engineers to be involved in the design process without overstressing the shared members of the team: the team lead and the top-level and system designers. There is also a natural support for hierarchy on large projects. Only two levels have been described in this chapter, but a large chip can be partitioned into major sections (ex. RF, analog, digital, etc.), with overall leaders for the whole chip, as well as leaders for the individual sections.

5.5 Supports IP Reuse

Not only does the top-down design process described in this document improve the communication between the members of the design team, but when the design is being reused, it also improves communication between design teams. If the design is being sold, then the process also improves the communications between different companies: seller and buyer.

A rigorous top-down design process creates as a natural by-product a thoroughly validated high-level description of the design, which is a critical enabler of IP reuse. This description is used by potential customers when evaluating the IP and by customers when integrating the IP. To see the value of having the needed model fall out of the design process as a by-product, consider the case where it does not. Using a bottom-up design process requires that the model be developed after the design is complete. This creates several barriers to the success of the IP. First, with the model not being used as an integral part of the design process it does not get much in the way of incidental testing. Substantial extra effort is required to field a high quality model, resulting in extra cost and delay. Furthermore, it is unlikely that the same quality model would be developed with an adjunct process. Second, with the model not being leveraged during the design process, the total cost of developing the model offsets any revenue from the IP, requiring higher levels of market success to break even. Finally, the model development process delays the release of the IP. This is especially troublesome as the price of IP drops dramatically as it becomes a commodity. Time-to-market is especially critical in the IP market as the price can drop by a factor of ten within a year of its release. Delay of even a month dramatically affects the total revenue of a product.

6 Final Words on Top-Down Design

Many design groups currently claim to be following a top-down design process, yet experience most of the problems attributed to the use of a bottom-up design style. This is because they are basically employing a bottom-up style with a few relatively cosmetic changes that serve to give the appearance of top-down design. This chapter lays out a series of principles that must be followed to realize all of the benefits associated with a rigorous top-down design methodology, with Verilog-AMS being the foundation upon which that methodology is built.

A rigorous top-down design methodology requires a significant investment in time and training and a serious commitment throughout the design process if it is to be successful. However, it is much easier the second time around and once mastered provides dramatic returns. Fewer design iterations and silicon re-spins are needed, which results in a shorter and more predictable design process. More optimal designs are produced that are better verified. It allows design teams to be larger and more dispersed, giving the option of trading a higher initial investment rate for a shorter time-to-market. And it is relatively tolerant of changes in the requirements that occur late in the design cycle.

Employing a rigorous top-down design methodology dramatically increases the effectiveness and productivity of a design team. If a design team fails to move to such a design style while its competitors do, it will become increasingly ineffective. It eventually will be unable to get products to market in a time of relevance and so will be forced out of the market.

Given the high pressure world that most designers live in, it is difficult for them to acquire the skills needed to be successful in a rigorous top-down design methodology. In addition, there is little training available from continuing education centers. This suggests that the transition to top-down design will be slow. The best hope for accelerating the move to top-down design is for universities to give designers the necessary background and training in the benefits and practice of rigorous top-down design. There are some signs that this is beginning [11], but it is not as aggressive or as widespread as it needs to be in order for there to be a smooth and timely transition.

What's Next

With Chapters 1-2 as a motivation as to the importance of Verilog-AMS, we are now ready to present the language itself. Verilog-A is introduced in the next chapter. With Verilog-A you can model purely analog components and blocks. Chapter 4 introduces Verilog-AMS. These two chapters should provide you with an understanding of the fundamental concepts of the two languages, but not all of the details. Those are presented in Chapter 5.

3 Analog Modeling

In this chapter, Verilog-A, the analog-only subset of Verilog-AMS, will be introduced using a series of practical examples, one example per section. In the beginning the examples will be simple, but they will be useful as is. As the chapter progresses the examples will become more advanced. Once the example is given, all aspects of it will be discussed. As new ideas are presented, they will be set in ***bold italics*** to make them easier to find and to call your attention to them as important points. Once an example is covered in detail, straight forward extensions to the concepts introduced by the example will be covered. Finally, pointers will be given to the language reference where more information can be found. These references appear like this (5§2.3p157), which includes the chapter number, the section number, and finally the page number. In this way, the language will be covered with a fair degree of completeness.

1 Resistor

One of the simplest models that can be described by Verilog-A is a resistor. In general, a resistor is a relationship between voltage and current, as in

$$f(v, i) = 0, \qquad (1)$$

where v represents the voltage across the resistor, i represents the current through the resistor, and f is an arbitrary function of two arguments. This is the most general definition of a resistor and so covers what people commonly refer to as a resistor (more precisely termed a linear resistor) as well as nonlinear resistors such as the intrinsic part of diodes and transistors. The resistance of a resistor is the derivative of the voltage with respect to current.

The equation for a simple linear resistor is

$$v = ri, \qquad (2)$$

where r is the resistance. A model for a linear resistor is given in Listing 1. This model uses only Verilog-A constructs and so can be used with both Verilog-A and Verilog-AMS simulators.

LISTING 1 *Verilog-A/MS description of a linear resistor.*

```
// Linear resistor (resistance formulation)

`include "disciplines.vams"

module resistor (p, n);
    parameter real r=0;    // resistance (Ohms)
    inout p, n;
    electrical p, n;

    analog
        V(p,n) <+ r * I(p,n);
endmodule
```

p, n, i, $+$ v $-$

$v = V(p,n)$
$i = I(p,n)$
$v = ri$

The first line of this model is

```
// Linear resistor (resistance formulation)
```

The // characters begin a ***comment*** (5§1.1p149), which extends to the end of the line. Comments are meant to explain the model to any person that might be trying to understand the model. They are ignored by whatever program is reading the model. In this book, comments will be placed in italics to make them easier to distinguish from the other parts of the model. Comments can also be written inline by using '/*' to start the comments, and '*/' to end them. Inline comments are rare, but this form is often use to write multi-line comments, such as

```
/*
 * RESISTOR
 * A linear resistor that uses the resistance formulation: v = ri
 */
```

Verilog-A/MS is a language that supports multiple disciplines. A ***discipline*** is a collection of related physical signal types, which in Verilog-A/MS are referred to as ***natures***. For example, the *electrical* discipline consists of voltages and currents, where both voltage and current are natures. Verilog-A/MS by itself defines only one discipline, the empty discipline, and it defines no natures. Thus, in order for the language to be able to describe models that operate on physical signals, the disciplines and natures associated with those signals must be defined. A collection of common disciplines and natures are defined in a file *disciplines.vams* (5§2.4p159) that is provided with all implementations of Verilog-A/MS. That file is included into this model by writing

```
`include "disciplines.vams"
```

The tick (`) that precedes the word *include* indicates this is a preprocessor directive (5§1.4p151). This line is replaced by the language preprocessor with the contents of the file *disciplines.vams* before being passed to the compiler. It defines the names *electrical*, *V*, and *I*, which are used later in the model. It also defines other disciplines and natures, but those are not used in this model. The name *include* is a keyword of the Verilog-A/MS language. Being a keyword, it is not a name that you can choose, both the name and its meaning are specified by the language itself (5§1.3p150). All keywords in listings are set in bold text.

It is not necessary to use *disciplines.vams*. You are free to create your own natures and disciplines. How to do so is described later in Section 3.1 on page 51.

The basic building blocks of Verilog-A/MS are ***modules***. Modules are descriptions of individual components (5§9.1p226). In Verilog-A/MS modules are a block of statements that begin with the keyword *module*, which is then followed by the name of the module and the list of ports. The statement is terminated with a semicolon.

```
module resistor (p, n);
```

A ***parameter*** is specified for the module using the *parameter* statement (5§2.3p157).

```
parameter real r=0;
```

In this case, a real valued parameter *r* is defined that can be specified when the module is instantiated (more about this later). The parameter is given a default value of 0, meaning that if the value is not specified when the module is instantiated, it will assume a value of 0. Thus with no value specified, the resistor will act as a perfect short circuit. All parameters must be given default values. However, specifying the type, in this case *real*, is optional. If not given, the parameter will take the type of the default value.

Ports are the points where connections can be made to the component (5§2.5p164). In this case, they are the terminals for the resistor. So far, the ports have only been given names, but have not been described in any other way. That is done in the two subsequent lines.

```
inout p, n;
electrical p, n;
```

These two lines describe the direction and the type of the ports. The ***port direction*** is given by the statement that begins with the keyword inout. There are three directions possible, input, output, and bidirectional as designated by the *input*, *output*, and *inout* keywords. Each port should be given a direction. Input ports can sense the signals that they are connected to, but cannot affect them; output ports can affect the signals, but cannot sense them; and inout ports can both sense and affect the signals. Since *inout* can do everything that both *input* and *output* ports can do, one might wonder why *input* and *output* ports are needed. In fact, they are not strictly needed. However, using input and output ports are considered a good practice because it provides clarity of intent. Labeling a port as either input or output at the top of the module makes the behavior of the module clearer. It also allows for extra error checking by whatever tool is reading the module.

The type of the ports is specified by the second of the two lines in which the name of a discipline is followed by a list of ports. In this case the *p* and *n* ports are defined to be *electrical*, meaning the signals associated with the ports are expected to be voltage and current.

The actual behavior of the module is defined in the next two lines.

```
analog
    V(p,n) <+ r * I(p,n);
```

The ***analog*** keyword introduces an ***analog process*** (5§6.1p196). An analog process is used to describe continuous time behavior. Syntactically, it is the analog keyword followed by a statement that describes the relationship between signals. This relationship must be true at all times. In this case, the statement that defines the relationship between the signals on the ports is a contribution statement. A ***contribution statement*** takes the form of a branch signal on the left side of a contribution operator, '<+', followed by an expression on the right side (5§3.2p169). The branch signal on the left side is forced to be equal to the value of the expression. The branch signal on the left is *V(p,n)*, it is the voltage across the implicit branch between the *p* and *n* ports. The expression on the right is *r∗I(p,n)*, the product of the parameter *r* and the branch signal *I(p,n)*, which is the current that is flowing through the implicit branch between the *p* and *n* ports. Thus, the contribution statement establishes a relationship between the branch voltage and the branch current that models a linear resistor with resistance *r*.

The signals *V(p,n)* and *I(p,n)* are the voltage across and the current through the ***implicit or unnamed branch*** between the nodes *p* and *n* (5§2.6p167). An implicit branch is referenced using its end points, in this case *p* and *n*. The signals associated with the branch are accessed using the ***access functions*** that are given in the definition of the discipline in *disciplines.vams* for the branch. An implicit branch inherits its discipline from its endpoints, both of which must have equivalent disciplines. In this case, the discipline of the end points *p* and *n* are *electrical*, and so the discipline of the implicit branch is *electrical*. The *electrical* discipline defines *V* as the access function for the potential of the branch and *I* as the access function for the flow through the branch. As such, the *V* in *V(p,n)* accesses the voltage across *p* and *n*, and the *I* in *I(p,n)* accesses the current that flows between *p* and *n*.

Finally the module definition is terminated with the **endmodule** keyword. Any statements that follow it are not associated with this module.

The resistor model given in Listing 1 is given using a resistance formulation, meaning that the voltage across the resistor is given as a function of the current through the resistor. It is also possible to use a conductance formulation, where the current through the resistor is given as a function of the voltage across the resistor. In this case the resistor is referred to as a conductor. Its constitutive relation is,

$$i = gv, \tag{3}$$

where g is the conductance. A model for it is given in Listing 2.

LISTING 2 *Verilog-A/MS description of a linear conductor.*

```
// Linear resistor (conductance formulation)
`include "disciplines.vams"

module conductor (p, n);
    parameter real g=0;  // conductance (Siemens)
    inout p, n;
    electrical p, n;

    analog
        I(p,n) <+ g * V(p,n);
endmodule
```

p, n, i, $+$ v $-$

$v = V(p,n)$
$i = I(p,n)$
$i = gv$

1.1 Capacitor

A capacitor is a relationship between voltage and charge, as in

$$f(v, q) = 0, \tag{4}$$

where v represents the voltage across the capacitor, q represents the charge through the capacitor, and f is an arbitrary function of two arguments. This is the most general definition of a capacitor. The capacitance is the derivative of the charge with respect to voltage. The equation for a simple linear capacitor is

$$q = cv, \tag{5}$$

where c is the capacitance. A model for a linear capacitor is given in Listing 3. This model uses only Verilog-A constructs and so can be used with both Verilog-A and Verilog-AMS simulators.

This model offers one complication not found in the earlier models. The model is formulated in terms of voltage and charge. We already know how to access the voltage on the capacitor, but not charge. There is no charge access function associated with *electrical* branches. However, as we have seen before, there is an access function for current, and current is related to charge in that

LISTING 3 *Verilog-A/MS description of a linear capacitor.*

```
// Linear capacitor
`include "disciplines.vams"
module capacitor (p, n);
    parameter real c=0;  // capacitance (F)
    inout p, n;
    electrical p, n;

    analog
        I(p,n) <+ c * ddt(V(p,n));
endmodule
```

$v = V(p,n)$

$i = I(p,n)$

$i = c\frac{dv}{dt}$

$$i = \frac{dq}{dt}. \tag{6}$$

Thus, the model is formulated in terms of voltage and current, and the charge is converted to current using the *ddt* operator (5§4.6.1p179), which returns the time derivative of its argument. Thus, the constitutive relation for the linear capacitor becomes,

$$i = c\frac{dv}{dt}, \tag{7}$$

which is encoded as a Verilog-A/MS contribution statement in Listing 3 as

```
I(p,n) <+ c * ddt(V(p,n));
```

1.2 Inductor

An inductor is a relationship between flux and current,

$$f(\phi, i) = 0, \tag{8}$$

where ϕ represents the flux across the inductor, i represents the current through the inductor, and f is an arbitrary function of two arguments. This is the most general definition of an inductor. The inductance is the derivative of the flux with respect to current. The equation for a simple linear inductor is

$$\phi = li, \tag{9}$$

where l is the inductance. A model for a linear inductor is given in Listing 4.

As with the capacitor, there is a complication. In this case, there is no access function for the flux of the branch, and so the flux is related to the voltage using

LISTING 4 *Verilog-A/MS description of a linear inductor.*

```
// Linear inductor
`include "disciplines.vams"

module inductor(p, n);
    parameter real l=0;    // inductance (H)
    inout p, n;
    electrical p, n;

    analog
        V(p,n) <+ l * ddt(I(p,n));
    endmodule
```


$$v = \frac{d\phi}{dt} \tag{10}$$

and the constitutive relation is reformulated in terms of voltage and current with

$$v = l\frac{di}{dt}. \tag{11}$$

1.3 Voltage and Current Sources

At this point, if you have understood what has been covered so far, you should be able to write Verilog-A models for an ideal DC voltage or current source. Please try to do so.

When finished, compare the models you wrote to those given in Listing 5 and Listing 6. Did you remember to change the direction of the ports from *inout* to *output*?

2 A Simple Circuit

Now that we have seen how to describe a component using a module, we will see how to combine instances of modules into a circuit. A resistor and a voltage source are combined in the circuit of Listing 7. The model starts off by including the models for the resistor (Listing 1) and the voltage source (Listing 5) using *include* statements. It is important to realize that these files, unlike *discipline.vams*, reside in the local directory, whereas *discipline.vams*, being a system file, resides in the installation directory. Generally simulators support a search path, which is an ordered list of directories that might contain include files. These directories are examined in turn by the *include*

LISTING 5 *Verilog-A/MS model for a constant-valued voltage source.*

```
// DC voltage source
`include "disciplines.vams"
module vsrc (p, n);
    parameter real dc=0;        // dc voltage (V)
    output p, n;
    electrical p, n;

    analog
        V(p,n) <+ dc;
endmodule
```

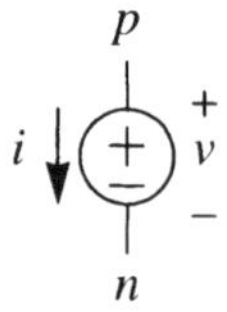

$v = V(p,n)$
$i = I(p,n)$
$v = v_{dc}$

LISTING 6 *Verilog-A/MS model for a constant-valued current source.*

```
// DC current source
`include "disciplines.vams"
module isrc (p, n);
    parameter real dc=0;        // dc current (A)
    output p, n;
    electrical p, n;

    analog
        I(p,n) <+ dc;
endmodule
```

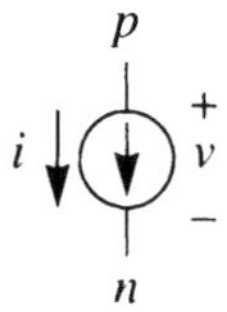

$v = V(p,n)$
$i = I(p,n)$
$i = i_{dc}$

statement when looking for files. It is usually possible to modify the search path, but the process of doing so is implementation specific. All simulators include the local directory and the installation directory on the search path.

The simple circuit of Listing 7 begins as did the others, with a comment and include of *disciplines.vams*, which provides the *electrical* discipline that are used to declare the nodes to be used in the circuit. Next, the previously given descriptions of the *vsrc* and *resistor* are included. Then comes the beginning of the module definition.

```
module smpl_ckt;
```

This differs somewhat from those already given in that the terminal declarations are missing. That is to be expected as this is the ***top-level circuit*** and as such has no terminals.

Next, two *electrical* nodes are defined: *n* and *gnd*. Nodes are used as interconnection points for ports. It is not possible to directly connect one port to another. Instead, one

LISTING 7 *Verilog-A/MS structural model.*

```
// A simple circuit
`include "disciplines.vams"
`include "vsrc.vams"
`include "resistor.vams"

module smpl_ckt;
    electrical n;
    ground gnd;

    vsrc #(.dc(1)) V1(n, gnd);
    resistor #(.r(1k)) R1(n, gnd);
endmodule
```

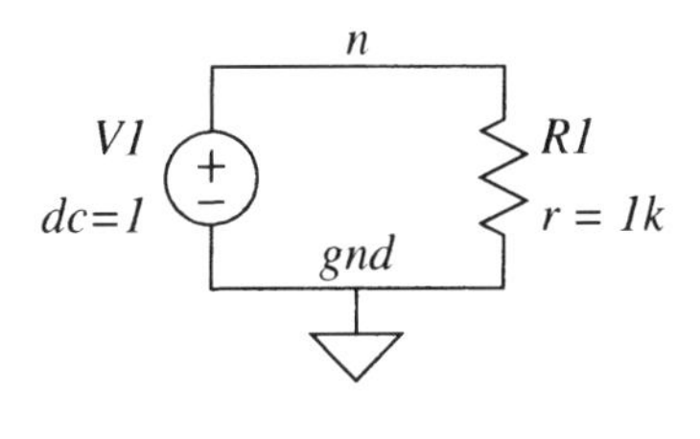

indirectly connects them by connecting the ports to the same node. Any number of ports can be connected to a node. Nodes are declared in a manner very similar to ports. The difference is nodes are not given in the port list in a module definition, and one does not specify the direction of nodes.

```
electrical n;
```

Every circuit has one node designated as the ***ground*** or ***reference node*** (5§2.5p164). This node defines zero potential for all disciplines, and as such it does not have a discipline itself. Local access to the reference node is provided through the *ground* statement.

```
ground gnd;
```

The ground statement defines a name, in this case *gnd*, that is used in the local context to refer to ground.

The circuit itself is constructed by creating instances of predefined modules, wiring them together by connecting them to nodes, and specifying parameters for them. This is done for the voltage source and the resistor in the following two lines.

```
vsrc #(.dc(1)) V1(n, gnd);
resistor #(.r(1k)) R1(n, gnd);
```

The first statement directs that an instance of the module *vsrc* be connected to nodes *n* and *gnd* and be named *V1*, and that its parameter *dc* be given a value of 1. The second directs that an instance of *resistor* be named *R1* and connect to *n* and *gnd*, and that its parameter *r* take a value of 1k. The value 1k is a number with a ***scale factor***, it is short-hand for 1000, and consists of a mantissa (a plain number without an exponent), in this case 1, followed by a scale factor, *k*, which is a symbol that means kilo or 1000.

The value used is the product of the mantissa and the scale factor (5§2.1.2p153). Table 2 on page 154 gives a complete list of scale factors.

In general, instance statements contain four things in the following order:

1. The name of the module being instantiated,
2. the list of values to be assigned to the module parameters (optional),
3. the name of the instance (optional), and
4. a list of nodes to which the modules ports are to be connected.

In this example, the two instance statements are partitioned into their component pieces in Table 1.

TABLE 1 *Partitioning of instance statements into their component pieces.*

Module Name	Parameter List	Instance Name	Node List
vsrc	#(.v(1))	V1	(n, gnd)
resistor	#(.r(1k))	R1	(n, gnd)

The module names and instance names are given in a straight forward manner. In the first case, you are specifying the names of existing module definitions, in the second you are assigning names for the instances you are creating. You are free to use any names that meet the basic Verilog naming conventions as long as the names are unique within the module.

The parameter list can be given as an unordered comma separated list of name value/ pairs as in Listing 7, or an ordered comma separated list of values without names. In either case, the parameter list is given within a matched pair of parentheses that follow a pound sign, as in

#(*parameter list*)

Name/value pairs are given with the following syntax

.*name*(*value*)

Assume a module has two parameters *p*1 and *p2* that were declared in that order. Further assume that values *v*1 and *v*2 should be assigned to parameter *p*1 and *p*2. This can be done using any of the following parameter lists (5§9.2p227),

```
#(.p1(v1), .p2(v2))
#(.p2(v2), .p1(v1))
#(v1, v2)
```

If one wanted to only specify a value for $p2$ and have $p1$ take its default value, use any of the following

```
#(.p2(v2))
#(.p1(), .p2(v2))
#(, v2)
```

Notice that when passing values to parameters by order, one simply leaves out the values for the parameters that should take their defaults. The commas are used to identify which values are missing.

The node lists can be specified in a similar manner (5§9.2p227). The only difference is that the leading pound sign is not used with node lists and the parentheses that surround the node list must be specified, even in the node list is empty. Thus, if a module supports two ports, $p1$ and $p2$, declared in that order, and they should be connected to nodes $n1$ and $n2$,

```
(.p1(n1), .p2(n2))
(.p2(n2), .p1(n1))
(n1, n2)
```

As with parameters, a node could be specified for $p2$ while having $p1$ left unconnected using any of

```
(.p2(n2))
(.p1(), .p2(n2))
(, n2)
```

Once all of the instances are specified, the module terminates with the *endmodule* statement. This type of module is often referred to as a ***structural module***, or a ***netlist***, as it consists only of instances of other modules. Structural modules are characterized as having instance statements, but no behavioral descriptions, such as the *analog* process. In contrast, the resistor module given in Listing 1 is considered a ***behavioral module***, as it contains an *analog* process and no instance statements. A behavioral model is simply a set of equations that have the property that the behavior of the signals at the terminals matches that of the component being modeled. It is important to recognize that the behavior of the quantities inside the model may not in any way match the behavior of the modeled component. This is part of the advantage of behavioral models; they can be abstractions that hide the complexity of the internal behavior of a component. Besides having structural and behavioral modules, it is also possible to have modules that contain both structure and behavior.

Any nodes that are not accessed in a behavioral block need not be declared. The reason for this is that within the module the signals on the node are not being accessed; only a connection is being made to the node. Thus, the module given in Listing 7 is longer than it needs to be. It can be shortened by removing the declaration of the nodes, as shown in Listing 8. In this case, the inclusion of *disciplines.vams* is removed because the definition of *electrical* is no longer needed, and the declaration of node *n* is removed. In addition, the inclusion of the resistor and voltage source definitions was removed. It is assumed that they are available in a library that is known to the simulator.

LISTING 8 *An abbreviated version of the module given in Listing 7.*

```
// A simple circuit
module smpl_ckt;
    ground gnd;

    vsrc #(.dc(1)) V1(n, gnd);
    resistor #(.r(1k)) R1(n, gnd);
endmodule
```

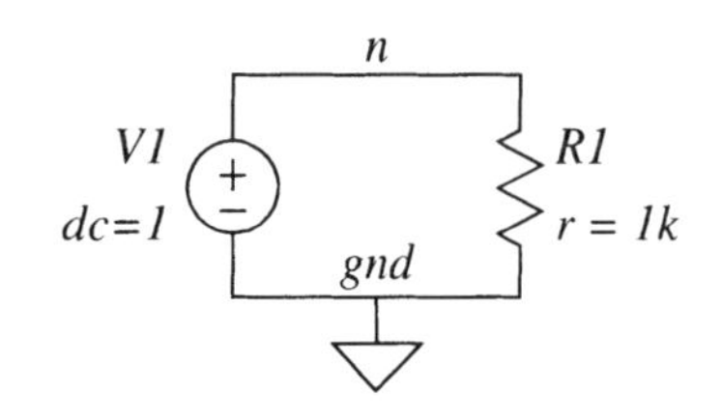

2.1 Conservative Systems

A ***system*** is considered to be a collection of interconnected ***components*** that are acted upon by a stimulus and produce a response. In Verilog-A/MS, components are instances of modules. The components themselves might also be systems, in which case a hierarchical system is defined. If a component does not have any sub-components, then it is considered a primitive component, which is necessarily a behavioral module. Each primitive component connects to one or more nodes through ports† or terminals as shown in Figure 1.

In order to simulate systems, it is necessary to have a complete description of the system and all of its components. Descriptions of systems are given structurally. That is, the description of a system contains instances of components and how they are interconnected. Descriptions of primitive components are given behaviorally. That is, a mathematical description is given that relates the signals at the ports of the component.

† The term ***port***, used heavily in the Verilog world to mean ***terminal*** or ***pin***, is somewhat of a misnomer. Port normally implies a pair of related terminals for which only two quantities are important, the port voltage and the port current. The port voltage is the potential difference between the two terminals, and the port current is the flow between the two terminals. The current into one terminal must exactly equal the current out of the other. Though this is the established definition of port, for consistency with Verilog terminology, in this book the term port is instead used to mean terminal.

FIGURE 1 *Components connect to nodes through ports.*

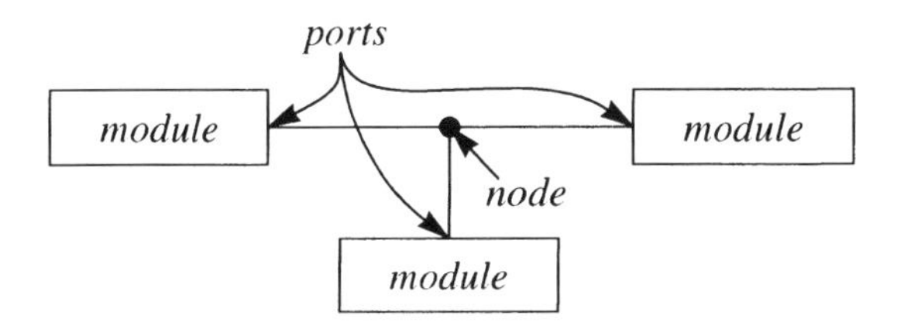

The circuit of Listing 7 is an example of a conservative system. A ***conservative system*** is a system where some quantity, such as energy or momentum, is explicitly conserved. By conserved it is meant, that when observed for the entire system, the conserved quantity must always be constant. Many types of systems can be decomposed and modeled in a relatively simple and compact manner using the concepts of conservative systems, particularly lumped systems that involve forces. Electrical networks, for example, are very naturally modeled as conservative systems, as are many mechanical and thermal problems. In general, anything that can be modeled with an 'electrical equivalent' can be modeled as a conservative system.

To support the modeling of conservative systems, Verilog-A/MS allows models to be formulated in terms of potentials and flows using the concepts of nodes and branches. A ***potential*** is a physical quantity that satisfies ***Kirchhoff's Potential Law*** (KPL). That is, when the quantity is accumulated around a closed path, it will always sum to zero. A ***flow*** is a physical quantity that satisfies ***Kirchhoff's Flow Law*** (KFL); that is when accumulated over a closed surface it must always total to zero. In electrical systems, voltage is an example of a potential, and both current and charge are examples of flows. KPL and KFL are generalizations of the famous Kirchhoff's Voltage Law (KVL) and Kirchhoff's Current Law (KCL). They are illustrated for a lumped network in Figure 2. Notice that the lumped network is drawn as a collection of ***nodes*** and ***branches***. A node is a point of interconnection for the branches, and a branch is a path between two nodes. As such, a branch always has two terminals and each terminal connects to one node. Both nodes and branches represent KFL surfaces, and as such, the total flow into either a node or branch must exactly equal the flow out of the same node or branch. Since a branch only has two terminals, the flow in one terminal must exactly equal the flow out of the other. Any closed loop of branches represents a KPL path, and as such, the potential on a branch exactly equals the difference in potential of the nodes to which it is attached. Any conservative lumped network can be represented as a collection of nodes and branches, though a single component may need to be represented by more than one branch. Furthermore, by Tellegen's Theorem [7] the branch potentials and flows satisfy

$$\sum_k P_k F_k = 0 \tag{12}$$

and so the network is conservative in the quantity that is the product of the potential and flow. Typically, that quantity is either energy or power, but as shown in Table 2, it need not be.

FIGURE 2 *Illustration of the concepts of potential and flow.*

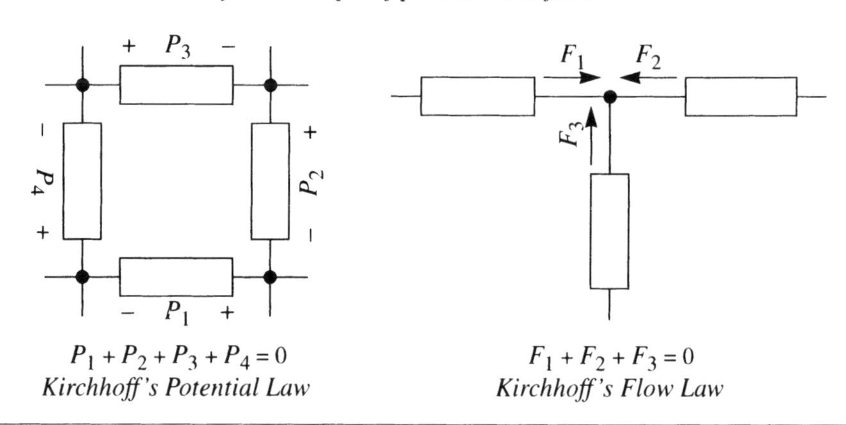

TABLE 2 *Some of the possible energy domains that can be used when formulating the models as conservative systems in Verilog-A/MS.*

Potential	Flow	Conserved
Electrical		
Electromotive Force (V)	Charge (C)	Energy (J)
Electromotive Force (V)	Current (A)	Power (W)
Magnetic		
Magnetomotive Force (A-turn)	Magnetic Flux (Wb)	Energy (J)
Magnetomotive Force (A-turn)	Magnetic Flux Rate (V)	Power (W)
Translational Kinematic		
Displacement (m)	Force (N)	Energy (J)
Velocity (m/s)	Force (N)	Power (W)

TABLE 2 *Some of the possible energy domains that can be used when formulating the models as conservative systems in Verilog-A/MS.*

Potential	Flow	Conserved
Rotational Kinematic		
Angle	Torque (N-m)	Energy (J)
Angular Velocity (/s)	Torque (N-m)	Power (W)
Thermal		
Temperature (K)	Entropy Flow (W/K)	Power (W)
Temperature (K)	Heat (J)	Energy-Temperature (J-K)
Temperature (K)	Heat Flow (J/s)	Power-Temperature (W-K)
Fluidic		
Pressure (N/m^2)	Flow (m^3)	Energy (J)
Pressure (N/m^2)	Flow Rate (m^3/s)	Power (W)
Radiant		
Luminous Intensity (cd)	Optical Flux (lm)	(cd^2-sr)

2.1.1 Reference Node

With conservative systems, Kirchhoff's Potential Law is formulated in terms of potential differences. This suggests that one could take a set of potentials that satisfy KPL for a network, and add a fixed value to each potential, and that the new set of potentials would also satisfy KPL. To eliminate this ambiguity, one node in the circuit is denoted the ***reference*** or ***ground node***. The potential on this node is fixed to be zero. In this way, the ambiguity is removed.

In a Verilog-A/MS system, the ground node always exists and one must simply connect to it. There are two ways of doing so. One way is to simply give the ground node a name and then use it when constructing a network. This approach is shown in Listing 7. In this example, the ground node is declared to be *gnd*, a name that is valid within the module that contains the declaration. Distinct modules may use different names for ground. For example, in one module it may be named *gnd*, in another *earth*, but both names refer to the same node. A different approach is typically used in behavioral models. In this case, one simply uses access functions with one argument.

In such cases, the other argument is assumed to be the ground node. For example, if the following equation is used to describe a floating resistor (with terminals p and n)

```
V(p,n) <+ r*I(p,n);
```

then the following would be used to describe a grounded resistor (with terminal p)

```
V(p) <+ r*I(p);
```

In this case, $V(p)$ accesses the potential from terminal p to ground, and $I(p)$ accesses the current that flows from terminal p to ground.

2.1.2 Reference directions

Verilog-A/MS assumes a particular set of reference directions for potentials and flows as shown in Figure 3. The reference direction for a potential is indicated by the plus and minus symbols near each terminal. The branch potential P is positive whenever the potential of the terminal marked with a plus sign (a) is larger than the potential of the terminal marked with a minus sign (b). Similarly, the flow F is positive whenever it moves in the direction of the arrow (in this case from + to –). For *electrical* branches, this potential is accessed using $V(a, b)$ and this flow is accessed using $I(a, b)$.

FIGURE 3 *Associated reference directions for potential and flow.*

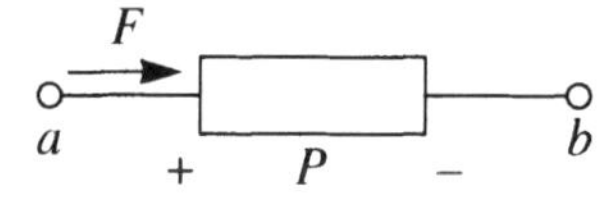

Verilog-A/MS uses associated reference directions [7]. A positive flow enters a branch through the terminal marked with the plus sign and exits the branch through the terminal marked with the minus sign.

3 Motor

An example of a multi-disciplinary model is a motor, which combines electrical and rotational kinematic modeling. A motor is described in rather simple terms by the following constitutive relations,

$$v = k_{\mathrm{m}}\omega + ri + l\frac{di}{dt}, \tag{13}$$

$$\tau = k_f i - d\omega - j\frac{d\omega}{dt}, \quad (14)$$

where v is the voltage applied across the winding in volts, i is the current through the winding in amperes, ω is the angular velocity of the shaft in radians per second, and τ is the torque acting on the shaft in Newton-meters. These equations are implemented in a Verilog-A/MS model given in Listing 9.

LISTING 9 *A motor model and a simple test bench.*

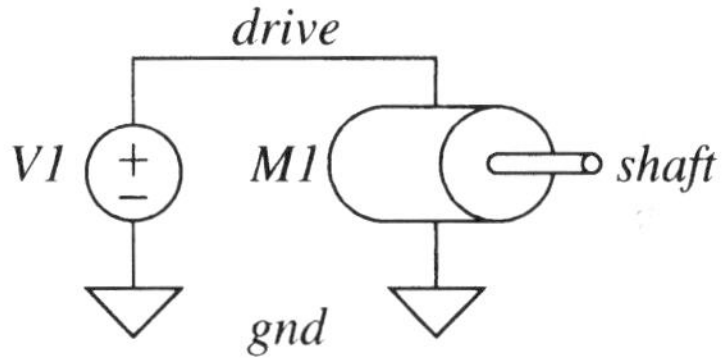

```
// A motor test circuit
`include "disciplines.vams"
`include "vsrc.vams"

module test;
    ground gnd;

    vsrc #(.dc(1)) V1 (drive, gnd);
    motor M1 (drive, gnd, shaft);
endmodule

// Motor
module motor(shaft, p, n);
    parameter real km = 4.5;      // motor constant (V-s/rad)
    parameter real kf = 6.2;      // flux constant (N-m/A)
    parameter real j = 0.004;     // inertia of shaft (N-m-s^2/rad)
    parameter real d = 0.1;       // drag (friction) (N-m-s/rad)
    parameter real r = 5.0;       // motor winding resistance (Ohms)
    parameter real l = 0.02;      // motor winding inductance (H)
    inout shaft, p, n;
    rotational_omega shaft;
    electrical p, n;

    analog begin
        V(p,n) <+ km*Omega(shaft) + r*I(p,n) + l*ddt(I(p,n));
        Tau(shaft) <+ kf*I(p,n) – d*Omega(shaft) – j*ddt(Omega(shaft));
    end
endmodule
```

drive
V1
M1
shaft
gnd

3.1 Natures and Disciplines

This model is constructed using two ***disciplines***, *electrical* and *rotational_omega*, which are shown in Listing 10 (5§2.4p159). Also shown are the natures used in these disciplines. These definitions are found in *disciplines.vams*.

LISTING 10 *The electrical and rotational_omega disciplines and their associated natures.*

```
// Voltage in volts
nature Voltage
    units = "V";
    access = V;
    abstol = 1u;
endnature

// Current in amperes
nature Current
    units = "A";
    access = I;
    abstol = 1p;
endnature

// Electrical
discipline electrical
    potential Voltage;
    flow Current;
enddiscipline
```

```
// Angular velocity in radians per second
nature Angular_Velocity
    units = "rads/s";
    access = Omega;
    abstol = 1u;
endnature

// Torque or Moment in Newton meters
nature Angular_Force
    units = "N-m";
    access = Tau;
    abstol = 1u;
endnature

// Rotational (velocity)
discipline rotational_omega
    potential Angular_Velocity;
    flow Angular_Force;
enddiscipline
```

A ***nature*** is used to describe basic physical quantities. The natures shown contain three pieces of information, the units of the quantity, a name used when accessing the quantity from a node, terminal, or branch, and a number that gives an indication of the expected size of the quantity. Essentially the nature combines these three pieces of information into a single unit and gives it a name, making it easy to incorporate them into a discipline as a group. Consider the nature for current,

```
nature Current
    units = "A";
    access = I;
    abstol = 1e–12;
endnature
```

The name of the nature is *Current*. By convention, the names for natures are always capitalized. This makes it easy to distinguish them from disciplines, which are always fully lower case. The units are given with the *units* keyword to be "A", which is short for amperes. The name of the access function is given with the *access* keyword to be "I", so the current on a branch b would be denoted $I(b)$. Finally, the scale of the signals of this nature is given with the *abstol* keyword to be 10^{-12} A, or 1 pA. The *abstol*, short for absolute tolerance, represents the largest amount of current that can always be considered negligible. It is used when simulating systems described in Verilog-A/

MS so that the simulator can trade simulation speed for accuracy. Without an understanding of what level of error is acceptable, the simulator would have to be so conservative that the simulations would not be practical.

Once defined, the natures can be incorporated into disciplines as a way of declaring the potential and flow. The disciplines are then used when declaring nodes, terminals, and branches.

Nature and discipline definitions are usually taken from the *disciplines.vams* header file (5§2.4p159), but you are also free to create your own. Often this occurs when modeling a type of system not currently supported by Verilog-A/MS, such as optical systems.

Natures include in their definition an indication of the scale of the signals they represent. But what if there are signals of the same underlying type that have vastly different scales? For example, consider the model of a high-voltage power distribution system being controlled by low-voltage electronics. The default definition for voltage and current sets the absolute tolerance to 1 μV and 1 pA respectively, values suitable for the low-voltage electronics, but much to small for the high-power sections of the system. One could define new voltage and current natures that have larger values for the absolute tolerance, but what is to be done with branches that span the two sections of the system. The concern is that Verilog-A/MS does not allow disciplines on each end of a branch to be incompatible. If the high- and low-power natures are defined independently, then the resulting high- and low-power disciplines will be incompatible. To avoid this problem, Verilog-A/MS provides the concept of ***base natures*** and ***derived natures***. Assume the nature defined for voltage and current in Listing 10 are the base natures, then deriving new natures for the high-power sections of the system would be done with

```
nature HighVoltage : Voltage
    abstol = 1;
endnature

nature HighCurrent : Current;
    abstol = 1e–3;
endnature

discipline hv_electrical
    potential HighVoltage;
    flow HighCurrent;
enddiscipline
```

In this case, *HighVoltage* is derived from *Voltage*, and inherits all of its properties except for *abstol*, which is set to 1 (any of the properties of the base nature can be overridden in the derived nature except the access function and the units). The same is true for *HighCurrent* and *Current*. Now, consider the following model fragment representing a resistor that spans the high and low voltage domains,

```
electrical lv;
hv_electrical hv;

analog
    I(hv,lv) <+ g*V(hv,lv);
```

This is legal because the *hv_electrical* and *electrical* disciplines are compatible. They are compatible because disciplines with compatible corresponding natures are compatible, and natures derived from the same base nature are compatible.

4 Junction Diode

An ideal diode is a component that allows current to flow in one direction but not in the other. A junction diode is a component that can be made using a semiconductor process that approximates this behavior. It is a nonlinear electrical component with the following characteristics,

$$i_{\mathrm{j}} = I_{\mathrm{s}}(e^{v/v_{\mathrm{T}}} - 1), \tag{15}$$

$$C_{\mathrm{j}} = \frac{C_{\mathrm{j}0}}{\sqrt{1 - \frac{v}{\phi}}}, \tag{16}$$

$$C_{\mathrm{d}} = \tau g_{\mathrm{d}} \text{ where } g_{\mathrm{d}} = \frac{di_{\mathrm{j}}}{dv} = \frac{I_{\mathrm{s}}}{V_{\mathrm{T}}} e^{v/v_{\mathrm{T}}} \approx \frac{i_{\mathrm{j}}}{V_{\mathrm{T}}}, \text{ and} \tag{17}$$

$$C_{\mathrm{t}} = C_{\mathrm{j}} + C_{\mathrm{d}}. \tag{18}$$

To implement the model in Verilog-A/MS it must be formulated as constitutive relations in terms of branch potentials and flows. This is already the case for the resistive portion of the model, (15), but not for the capacitive part, (18). To avoid the charge conservation problems the constitutive relationship of the nonlinear capacitor must be formulated in terms of charge and voltage [19,20]. To do so, the capacitance of (18) is integrated with respect to voltage to find the charge,

$$q_{\mathrm{c}} = \tau_{\mathrm{f}} i_{\mathrm{j}} - 2C_{\mathrm{j}0}\phi\sqrt{1 - \frac{v}{\phi}}. \tag{19}$$

Then the total diode current results from combining (15) and (19),

$$i_{\mathrm{d}} = i_{\mathrm{j}} + \frac{dq_{\mathrm{c}}}{dt}. \tag{20}$$

The Verilog-A/MS model that implements these equations is given in Listing 11.

LISTING 11 *Verilog-A/MS model for a junction diode (this model should not be used in practice as it fails when $v > \phi$).*

```
// Junction diode
`include "disciplines.vams"
module diode (a, c);
    parameter real is=10f from (0:inf);     // saturation current (A)
    parameter real tf=0 from [0:inf);       // forward transit time (s)
    parameter real cjo=0 from [0:inf);      // zero-bias junction capacitance (F)
    parameter real phi=0.7 exclude 0;       // built-in junction potential (V)
    inout a, c;
    electrical a, c;
    branch (a, c) res, cap;
    real qd;

    analog begin
        I(res) <+ is*(limexp(V(res)/$vt) – 1);
        qd = tf*I(res) – 2*cjo*phi*sqrt(1 – V(cap)/phi);
        I(cap) <+ ddt(qd);
    end
endmodule
```

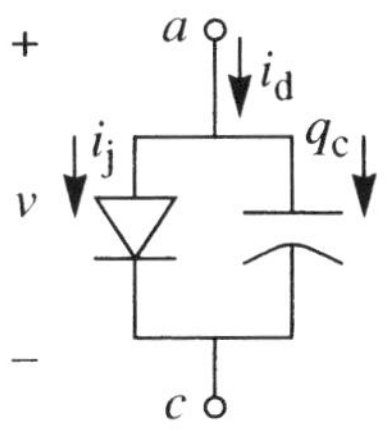

The model begins with the traditional inclusion of the disciplines. The module itself is given the name *diode* and two *electrical* terminals are named *a* and *c*, which represent the anode and cathode.

Four parameters are declared, *is*, *tf*, *cjo*, and *phi*, which represent i_s, τ_f, C_{j0}, and ϕ.

```
parameter real is=10f from (0:inf);     // saturation current (A)
parameter real tf=0 from [0:inf);       // forward transit time (s)
parameter real cjo=0 from [0:inf);      // zero-bias junction capacitance (F)
parameter real phi=0.7 exclude 0;       // built-in junction potential (V)
```

New in these declarations is the use of ***range limits*** to constrain the values specified for these parameters to a particular range (5§2.3p157). Range limits, when given, always follow the default value in a parameter declaration or another range limit, and take the form of a keyword (either *from* or *exclude*) followed by a range. Use of the *from* keyword indicates that the parameter value must be within the range, the *exclude* keyword indicates that it must not. The range may either be a point or an interval. The declaration of *phi* contains an example of a point exclusion range.

The other range limits used in the diode model are interval limits. An interval is defined by giving its two end points within delimiters and separated by a colon. The delimiters may either be parentheses or brackets and need not be matched. Parentheses are used to indicate that the end of the interval is open and brackets indicate the end is closed. A closed end includes the end point itself, whereas an open end does not. To specify that the parameter has no bound on one end, the end point is given to either be *–inf* if it is the left end point, or *inf* if it is on the right. The keyword *inf* is used to designate infinity.

TABLE 3 *The meanings of various intervals used in range limits. The parameter value is denoted with x; a and b represent the upper and lower limits of the interval.*

[a,b]	$a \le x \le b$	[a,**inf**)	$a \le x$
(a,b]	$a < x \le b$	(a,**inf**)	$a < x$
[a,b)	$a \le x < b$	(–**inf**,b]	$x \le b$
(a,b)	$a < x < b$	(–**inf**,b)	$x < b$

From the parameter declarations given in Listing 11, it should be clear that the value specified for *is* must be positive ($0 < is$), the value for *tf* and *cjo* must be nonnegative ($0 \le tf$, $0 \le cjo$), and *phi* must not be zero. In general it is a good practice to include range limits on all parameters for which it is appropriate as it improves the robustness of your model and makes your model easier to understand. Also, adding a description to your parameter as done in this model is a great habit to get into, as is adding the units to the description if appropriate.

A new type of statement is given next. A ***branch declaration statement*** is used to define ***explicit*** or ***named branches*** (5§2.6p167).

```
branch (a, c) res, cap;
```

In this case two branches are created, both of which are connected between ports *a* and *c*. As such, these two branches are in parallel. The discipline for the branches is *electrical*, because the end point *a* and *c* are declared to be *electrical*. The disciplines on the end points of a branch must always be compatible, a constraint that was described more fully in Section 3.1 on page 51. One of these branches, *res*, will be used to represent the resistive part of the diode model, and the other, *cap*, will represent the capacitive part. By declaring two parallel branches, it is possible to distinguish the currents flowing through the two branches. This would not be possible using the implicitly declared branches used earlier, which only identifies a branch by its endpoints.

The last declaration given in Listing 11 is

```
real qd;
```

This simply declares *qd* to be a ***real variable*** (5§2.2p155). Real variables are always initialized to zero. Verilog-A/MS also allows integer variables to be declared. Each variable, whether it be integer or real, is associated with a kernel, either continuous time or discrete event, depending on where the variable is assigned a value. If assigned a value in an analog process, it is associated with the continuous time kernel; otherwise it is associated with the discrete event kernel. Integer variables associated with the continuous time kernel are initialized to 0; those associated with the discrete event kernel are initialized to *x*, or unknown.

The rest of the model contains the constitutive equations for the model. The equations are given in the form of two contribution statements within an analog process. In this case, the analog process contains more than a single statement, and so the beginning and end of the process are delineated by the *begin* and *end* keywords. The first contribution statement gives the branch equation for the resistive part of the diode (5§3.2p169).

```
I(res) <+ is*(limexp(V(res)/$vt) – 1);
```

This statement computes the branch current I(res) from the branch voltage V(res), and so implements (15). It employs two unique capabilities of Verilog-A/MS, the **limexp**() function and $vt system function. The '$' symbol is used before a name to denote a system task or system function. The $vt system function returns the thermal voltage, *kT/q*, for the current temperature. $vt takes an optional argument, the temperature in Kelvin; when given it will return the thermal voltage associated with that temperature. Without an argument, $vt is equivalent to $vt($temperature), where $temperature is a system function that returns the current ambient temperature (5§4.5p175).

The **limexp**() function is functionally identical to the **exp**() function. When given an argument of x both return e^x. However, **limexp**() also performs limiting, and hence its name, which is short for *limited exponential* (5§4.8.1p188).

Limiting is a technique used to help improve the convergence of the simulator running the model. An exponential is a very nonlinear function, which if given a large argument will return an extremely large value. This can occasionally cause problems for the iterative methods used within simulators, causing them to fail because they never converge. Limiting places a bound on how fast the return value can change from iteration to iteration. This bound is actually a complex function of the current and past arguments and is not something you can set. The **limexp**() function coordinates with the simulator to prevent it from declaring convergence while the function is limiting, so externally the **limexp**() appears identical to **exp**() because you can never observe the intermediate, limited, values. The **limexp**() function was provided in order to model the *IV* characteristics of *pn* junctions, and generally should only be used in that application as the **limexp**() function is a more expensive function that **exp**() both in terms of time and memory.

The constitutive equation for the diode capacitors is given by

```
qd = tf*I(res) – 2*cjo*phi*sqrt(1 – V(cap)/phi);
I(cap) <+ ddt(qd);
```

which implements (19). It is interesting to note that the charge associated with the diffusion capacitance, $\tau_f i_j$, is a function of the branch current through the resistive part of the model, I(res). This is the reason why the model was implemented with two branches, it was important to distinguish this current from the total current. It is also worth noting that the order in which signals on nodes, terminals, and branches are accessed is of no consequence. In this model, a current is contributed to I(res) before it is used to determine the charge in the diffusion capacitance, but it need not be. The model could have been rewritten as

```
qd = tf*I(res) – 2*cjo*phi*sqrt(1 – V(cap)/phi);
I(cap) <+ ddt(qd);
I(res) <+ is*(limexp(V(res)/$vt) – 1);
```

without changing the results at all. In effect, the final value of I(res), or any signal with continuous-time discipline, is known at the beginning of the analog process and its value does not change during the evaluation of the process. Furthermore, its value exactly matches the total amount contributed to it during the evaluation of the analog process. It is as if the signal is psychic, knowing what its final value will be before evaluation begins. How does it manage this amazing feat? It's simple really. The analog process is evaluated iteratively along with all of the analog processes in the system, and iteration continues until all of the signals converge to values that satisfy all of the constitutive equations.

This idea is exploited in the next section to add a series parasitic resistor to the resistive part of the diode.

4.1 Junction Diode with Series Resistor

Consider stripping the capacitance from the model of Listing 11 and replacing it with a series resistor. The result is given in Listing 12, which implements

$$i = I_s(e^{(v - ir)/v_T} - 1) . \tag{21}$$

LISTING 12 *Verilog-A/MS model for a simple diode with a series resistance.*

```
// Junction diode
`include "disciplines.vams"
module diode (a, c);
    parameter real is=10f from (0:inf);      // saturation current (A)
    parameter real r=0 from [0:inf);         // series resistance (Ohms)
    inout a, c;
    electrical a, c;

    analog
        I(a,c) <+ is*(limexp((V(a,c) – r*I(a,c))/$vt) – 1);
endmodule
```


This equation is very similar to (15), with the exception that the current i is found on both sides of the equal sign. As such, this is an ***implicit equation***. In this sense it is different in an important way from (15), which is an explicit equation or formula. In (15) one can determine the value of i simply by evaluating the right hand side of the equation. With (21), that is not possible.

At this point it is important to note that the symbol '=' is used in this book in two different ways. When found in mathematical equations, like (21), it is used to represent ***equality***, that the value of the expression on its left must be the same as the value of the expression on its right, with no inherent suggestion about how to make them so. An equation is simply a statement of fact, not a recipe. When used in Verilog-A/MS code, '=' represents an ***assignment***, and so the value of the expression on its right side replaces the value of the variable on its left. Consider the equation,

$$x = 2x. \tag{22}$$

This equation can be solved by subtracting x from both sides to find that $x = 0$. Now consider the following fragment of Verilog-A/MS code,

```
x = 2x;
```

In this case, the value of x after the statement cannot be known without knowing what it was before the statement was evaluated. Say that initially the value of x is 5; after the assignment statement is evaluated, the value will be 10. Notice that equations are statements of fact and the order in which they are given is of no consequence, whereas assignments are operations, and the final result depends on the order in which they are performed. For example, after

```
x = 2;
x = 2x;
```

the value of x is 4, whereas x would be 2 if these statements were reversed.

In Verilog-A/MS it is the contribution statements that are very much like equations (contributions use '<+' rather than '='). As such, (21) can be written

```
I(a,c) <+ is*(limexp((V(a,c) – r*I(a,c))/$vt) – 1);
```

Notice that I(a,c) is found on both sides of the contribution operator. The underlying semantics of the language require that a value of I(a,c) be found such that

1. The value of I(a,c) used in the expression on the right side of the contribution operator is the same as the resolved value of I(a,c) (after all contributions have been made to it).
2. The value of the expression on the right side of the contribution operator is the same as the value of the resolved signal on the left side, in this case I(a,c), (again, after all contributions have been made to it).

Contributions have three characteristics not associated with equations. First, the left side of a contribution must be a branch signal. Second, if there are multiple contributions being made to the same branch, the contributions sum. This feature can be used to make models more modular. For example, if it is desirable to add a leakage to the diode, it can be easily done simply by appending the following line to the end of the analog process,

```
I(a,c) <+ gleak*V(a,c);
```

Finally, there is directionality associated with the contribution statement that is not present with equations. This distinction is discussed next.

4.2 Probes and Sources

Listing 13 contains a module that implements a voltage-controlled voltage source (VCVS). The equation for a voltage-controlled voltage source is

$$v_{\text{out}} = \alpha v_{\text{in}}, \tag{23}$$

where α represents the gain of the source. But this is not a complete description of a VCVS. There is directionality to a controlled source that is not captured by this equation alone. What is missing is the equation,

$$i_{\text{in}} = 0. \tag{24}$$

However, this equation is not found in Listing 13. It is not needed because contributions in Verilog-A/MS innately incorporate directionality. It does this by using the concept of probes and sources.

LISTING 13 *Verilog-A/MS model for a linear voltage-controlled voltage source.*

```
// Voltage-controlled voltage source
`include "disciplines.vams"

module vcvs (p, n, ps, ns);
    parameter real gain=1;     // voltage gain (V/V)
    output p, n;
    input ps, ns;
    electrical p, n, ps, ns;

    analog
        V(p,n) <+ gain*V(ps,ns);
endmodule
```

v_{in} v_{out}

$v_{out} = \alpha v_{in}$

In Verilog-A/MS, any signal that is the target of a contribution statement can conceptually be considered a source, and any signal used in an expression can be considered a probe. With the VCVS, both the driven and observed signals are voltages, so the source is a voltage source and the probe is a voltage probe, as shown in the figure inset in Listing 13.

If one understands the directional nature of contribution statements, thinking of the behavioral description as a set of related probes and sources is not necessary. However, it does allow one to interpret any analog behavioral description as a network, and many people find that helpful. To do so, there are a few rules that must be understood.

Probes. Any branch, be it explicit or implicit (named or unnamed), is a ***probe branch*** if there is no contribution made to that branch. There are two types of probe branches, potential probes and flow probes, as shown in Figure 4. A probe branch is a ***flow probe*** if its flow is observed somewhere in the module, otherwise it is a ***potential probe***. The potential across a flow probe is zero and the flow through a potential probe is zero. It is not possible for a probe to simultaneously be a potential and flow probe, and so it is illegal to observe both the potential and flow of a probe branch.

FIGURE 4 *Equivalent circuit models for the probe branches.*

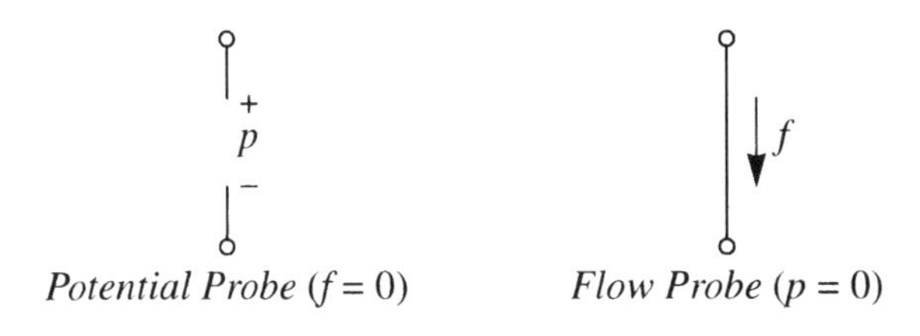

Sources. Any branch, either explicit or implicit, is a ***source branch*** if either the potential or the flow of that branch is assigned a value by a contribution statement anywhere in the module. It is a ***potential source*** if the branch potential is specified and is a ***flow source*** if the branch flow is specified. A branch cannot simultaneously be both a potential and a flow source, although it can switch between them, in which case it becomes a ***switch branch***. To switch a branch to being a potential source, assign to its potential. To switch a branch to being a flow source, assign to its flow. This type of branch is useful when modeling ideal switches, ideal diodes, mechanical stops, etc. Both the potential and the flow of a source branch are accessible for observation (they can be used in an expression). The models for the various branches are shown in Figure 5.

FIGURE 5 *Equivalent circuit models for the source branches.*

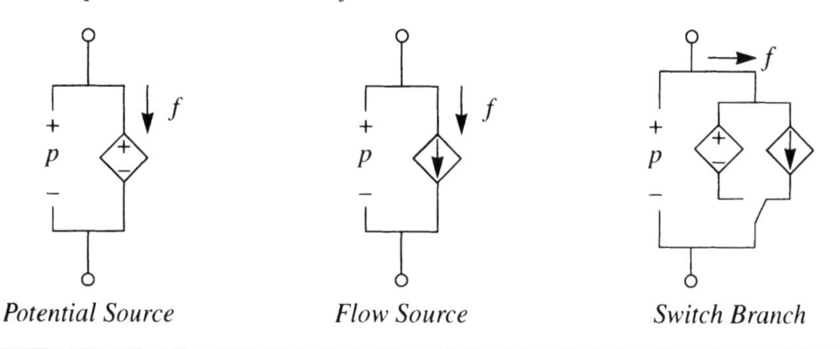

It is interesting to note that a potential probe is essentially a flow source where the contributions to the flow total to zero and a flow probe is essentially a potential source where the contributions to the potential total to zero.

Examples. The VCVS can be converted to any of the three other basic controlled sources by replacing

```
V(p,n) <+ gain*V(ps,ns);
```

in Listing 13 with one of the following:

Voltage-Controlled Current Source (VCCS)

```
I(p,n) <+ gain*V(ps,ns);
```

Current-Controlled Voltage Source (CCVS)

```
V(p,n) <+ gain*I(ps,ns);
```

Current-Controlled Current Source (CCCS)

```
I(p,n) <+ gain*I(ps,ns);
```

These probes and sources can be combined to model most any type of component. For example, consider the resistor of Listing 1 on page 36, from which, the contribution statement is

```
V(p,n) <+ r * I(p,n);
```

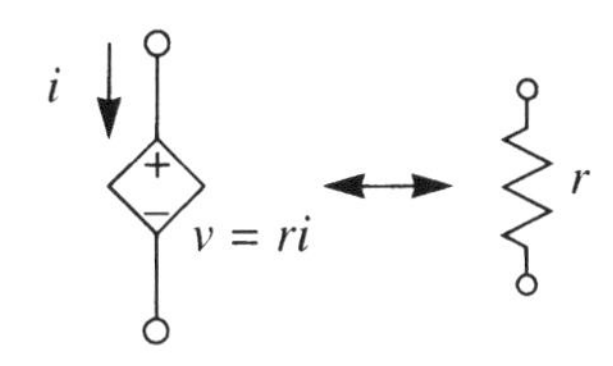

The contribution is made to the potential of the branch, making the branch a potential source branch. The use of I(p,n) in the expression on the right side then simply accesses the flow through this potential source branch.

The situation is similar for the conductor of Listing 2 on page 39. Here the contribution statement is

```
I(p,n) <+ g * V(p,n);
```

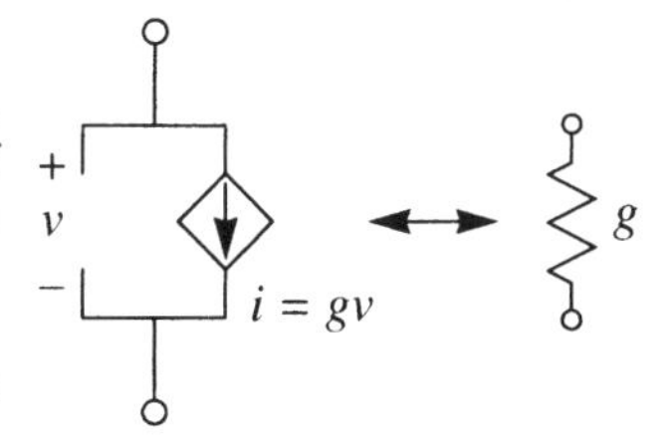

The contribution is made to the flow of the branch, making the branch a flow source branch. The use of V(p,n) in the expression on the right side then simply accesses the potential through this flow source branch.

An example that involves a switch branch is given in Section 6.

4.3 Series and Parallel RLC

As an exercise, try to write models for both a series RLC and a parallel RLC using a single behavioral model for each. Write two versions of each model, one that uses named branches, and one that does not. The constitutive equation for a series RLC is

$$v = v_R + v_L + v_C = ri + l\frac{di}{dt} + \frac{1}{c}\int idt \tag{25}$$

and the constitutive equation for the parallel RLC is

$$i = i_R + i_C + i_L = \frac{v}{r} + c\frac{dv}{dt} + \frac{1}{l}\int vdt\,. \tag{26}$$

One hint, the integration operator in Verilog-A/MS is *idt* (5§4.6.2p179).

When finished, compare the models you wrote to those given in Listing 14 and Listing 15. Did you remember to place limits on the parameter values to avoid any possible case of division by zero? Notice how the use of unnamed branches constrained the formulation of the equations for the individual components, which ultimately resulted in 0 being excluded from the range of several parameters.

LISTING 14 *Verilog-A/MS models for a linear series RLC.*

```
// Series RLC
`include "disciplines.vams"
module series_rlc (p, n);
    parameter real r=0;
    parameter real l=0;
    parameter real c=1p exclude 0;
    inout p, n;
    electrical p, n;
    analog begin
        V(p,n) <+ r*I(p,n);
        V(p,n) <+ l*ddt(I(p,n));
        V(p,n) <+ idt(I(p,n))/c;
    end
endmodule
```

```
// Series RLC
`include "disciplines.vams"
module series_rlc (p, n);
    parameter real r=0;
    parameter real l=0;
    parameter real c=0;
    inout p, n;
    electrical p, n, i;
    branch (p, i) rl, (i, n) cap;
    analog begin
        V(rl) <+ r*I(rl);
        V(rl) <+ l*ddt(I(rl));
        I(cap) <+ c*ddt(V(cap));
    end
endmodule
```

LISTING 15 *Verilog-A/MS models for a linear shunt RLC.*

```
// Shunt RLC
`include "disciplines.vams"
module shunt_rlc (p, n);
    parameter real r=1 exclude 0;
    parameter real l=1n exclude 0;
    parameter real c=0;
    inout p, n;
    electrical p, n;
    analog begin
        I(p,n) <+ V(p,n)/r;
        I(p,n) <+ c*ddt(V(p,n));
        I(p,n) <+ idt(V(p,n))/l;
    end
endmodule
```

```
// Shunt RLC
`include "disciplines.vams"
module shunt_rlc (p, n);
    parameter real r=0;
    parameter real l=0;
    parameter real c=0;
    inout p, n;
    electrical p, n;
    branch (p, n) res, cap, ind;
    analog begin
        V(res) <+ r*I(res);
        I(cap) <+ c*ddt(V(cap));
        V(ind) <+ l*ddt(I(ind));
    end
endmodule
```

5 Resistive Port

A resistive port is a component that is used to interface to high frequency circuits. Do not confuse it with the name that Verilog-A/MS uses to signify a module terminal. A resistive port combines a voltage source and a resistor, and so can be used as either a resistive source or a load. The constitutive equations for the resistive port are

$$v = 2v_{\mathrm{dc}} + ri, \qquad (27)$$

where v is the voltage across the port and i is the current through the port. The factor of 2 in this equation is present to assure that the desired output voltage is achieved when the port is terminated in its characteristic resistance, r. The resistive port model is given in Listing 16. This model starts off by including a new file, *constants.vams*, which contains preprocessor directives that define useful mathematical and physical constants. In this case, the value of Boltzmann's constant, k, is needed, which is accessed later using `P_K. In this file, names that start with `P_ represent physical constants and those that start with `M_ represent mathematical constants (5§2.1p152).

LISTING 16 *Verilog-A/MS description of a resistive port.*

```
// Resistive port

`include "disciplines.vams"
`include "constants.vams"

module port (p, n);
    parameter real r=0 from (0:inf);     // port resistance (Ohm)
    parameter real dc=0;                 // DC level into a matched load (V)
    parameter real mag=0;                // AC magnitude into a matched load (V)
    inout p, n;
    electrical p, n;

    analog begin
        V(p,n) <+ r*I(p,n) + 2*dc;
        V(p,n) <+ 2*ac_stim(, mag);
        V(p,n) <+ white_noise(4*`P_K*$temperature*r, "thermal");
    end
endmodule
```

This model adds support for ***small-signal analysis***. A small-signal analysis is a particular form of simulation where the system is first linearized about some operating point, and then the response to some 'small' stimulus is determined. Here the term small implies that the stimulus is so small that it is unable to excite any nonlinear behavior. This is the assumption that allows the circuit to be linearized to start the analysis. The linearization is performed on all models automatically by the simulator. However, the act of linearization eliminates all stimuli (adding a stimulus is a nonlinear process). As a result, the small-signal stimulus must be explicitly included in the models that are to act as small-signal sources. Verilog-A/MS provides special functions for this (5§4.9p189).

There are two kinds of small-signal analyses, AC and noise. In an AC analysis, a small number of small-signal sources (usually one) are explicitly placed in the circuit to allow the transfer functions from that source to the outputs to be determined. In a noise analysis, the stimuli are the internal noise sources that are inherent to most components (thermal noise, shot noise, flicker noise, etc.). The port is configured both to act as a small-signal stimulus for AC analyses, and as a noise source for noise analyses. The *ac_stim* function takes three arguments and returns 0 except during an AC analysis whose name matches *analysisName*, when it returns a signal with the specified magnitude and phase.

ac_stim(*analysisName*, *magnitude*, *phase*)

The default analysis name is "ac", magnitude is 1, and phase is 0. In this model, the default analysis name is used (notice that the first argument is missing), and the magnitude is specified by the module's *ac* parameter.

```
V(p,n) <+ ac_stim(, ac);
```

For the noise analysis, Verilog-A/MS provides three different noise stimulus functions, *white_noise*, *flicker_noise*, and *noise_table*. The *white_noise* function produces noise whose power is independent of frequency. The *flicker_noise* function produces noise whose power varies with frequency as $1/f^a$, where a is a constant. Finally, *noise_table* produces noise whose power varies as a piecewise function of frequency.

white_noise(*pwr, name*)
flicker_noise(*pwr, a, name*)
noise_table(*array, name*)

The desired noise power is passed into these functions using the first argument. In the first two the power is a simple real scalar. In *noise_table*, one passes in the power as a function of frequency by providing a vector of real pairs, the first number in each pair is the frequency and the second is the power. The last argument is the source name. It is used to identify the noise source in a noise results summary when there are multiple noise sources in the same instance. Finally, the second argument for *flicker_noise* is a, the negative of the exponent of frequency.

The port exhibits thermal noise due to its resistive component:

$$\bar{v}^2 = 4kTr. \tag{28}$$

It is modeled with

```
V(p,n) <+ white_noise(4*`P_K*$temperature*r, "thermal");
```

As with *ac_stim*, the *white_noise* function returns 0 for the analyses where it is not active. During a noise analysis, the statement produces noise with a power of $4kTr$ and a tag of "thermal".

As an exercise, try adding thermal noise to the resistor of Listing 1 on page 36.

6 Relay

The model for an ideal relay is shown in Listing 17. A relay is an electronically controlled switch. This relay is ideal in the sense that when the relay is closed, there is no voltage across its contacts, and when it is open there is no current flowing though its contacts.

LISTING 17 *Verilog-A/MS model for an ideal relay.*

```
// Ideal relay
`include "disciplines.vams"

module relay (p, n, ps, ns);
    parameter real thresh=0;  // threshold (V)
    output p, n;              // the contacts
    input ps, ns;             // the coil
    electrical p, n, ps, ns;

    analog begin
        @(cross( V(ps,ns) – thresh, 0 ))
            ;
        if (V(ps,ns) > thresh)
            V(p,n) <+ 0;
        else
            I(p,n) <+ 0;
    end
endmodule
```

The analog process begins with an event statement (5§6.8p204). In this statement the ***cross*** function creates an event when the value of its first argument goes through zero in the direction specified by the second argument. The second argument, being zero, indicates that events should occur if the crossing occurs in either direction. The other choices are +1 and –1, meaning that events should occur only on positive and negative going crossings respectively. Any other value prevents the events from being generated at all.

The event statement has the following form,

> **@**(*event expression*)
> *event clause (executed when events occur)*

The event that is generated by the *cross* function is caught by the @ statement and causes the event clause (or the body of the event statement) to be executed. In this case, the event clause is empty, and so it would appear as if the event statement is not affecting the behavior of the model. However, it is important to realize that besides returning an event when it detects a valid threshold crossing, the *cross* function also causes the simulator to place an evaluation point very close (and just after) the threshold crossing. In this model, that has the effect of causing the relay to carefully resolve the time when it opens and closes.

It is relatively common in analog processes to have event statements with empty event clauses. Notice that when this occurred in this module, the semicolon that terminates the event statement was placed on the line below the event expression and indented. This is a good practice because it makes it obvious that the event statement has an empty event clause. Contrast the way the analog process was formatted in Listing 17 to the way it is formatted below.

```
analog begin
    @(cross( V(ps,ns) – thresh, 0 ));
    if (V(ps,ns) > thresh)
        V(p,n) <+ 0;
    else
        I(p,n) <+ 0;
end
```

A casual glance at this version leads one to think that the *if* statement is the event clause. This version is somewhat misleading and so should not be used.

The relay model also introduces *if* statements or conditionals, and demonstrates the use of a switch branch. There are two forms of the *if* statement,

> **if** (*condition*)
> *if clause (executed if condition is true)*

or

> **if** (*condition*)
> *if clause (executed if condition is true)*
> **else**
> *else clause (executed if condition is false)*

The condition is an expression that can evaluate either to an integer or a logical value. If the result is true, or nonzero, then the *if* clause is executed. If the result is false, or zero, then the *else* clause is executed if it exists. If these clauses consist of more than one statement, they should be embedded between the *begin* and *end* keywords. In this example, the *if* clause contains a contribution statement that sets the voltage difference between the contacts, *p* and *n*, to zero and the *else* clause sets the current flow through the contacts to zero, making (*p*,*n*) a switch branch. If a potential is contributed to a branch at times, and at other times nothing is contributed, the branch is still considered a switch branch. As such, the analog process of the ideal relay could be shortened to

```
analog begin
    @(cross( V(ps,ns) – thresh, 0 ))
        ;
    if (V(ps,ns) > thresh)
        V(p,n) <+ 0;
end
```

However, the converse is not true. If a flow is contributed to a branch at times, and at other times nothing is contributed, the branch is considered a flow branch.

6.1 Non-Ideal Relay

The model for a non-ideal relay is shown in Listing 18. The relay is non-ideal in it exhibits both an on resistance and an off resistance.

This model differs from the one in Listing 17 in that the on and off resistance are passed as parameters. Notice that the model is written in such a way that the on resistance may be zero and the off resistance may be infinite, and that this is the case by default. So, by default, this model is ideal. This is possible because the contacts are modeled using a switch branch.

In Listing 19 a non-ideal relay is modeled without using a switch branch. In this case the contact resistance is not allowed to be zero so the relay cannot be ideal.

Notice the call to the *discontinuity* function in the body of the @ block (5§5.1.2p191). This function is used to announce discontinuities in the model to the simulator. In this case, it is announcing the discontinuity that occurs in the resistance of the contacts when the contacts open or close (when the input voltage crosses the threshold in either direction). The argument is the order of the discontinuity, with a value of *i* representing a discontinuity in the ith derivative of the constitutive relations for the component with respect to either a signal value or time. Thus, passing a 0 implies a break in the model and passing a value of 1 implies a kink.

LISTING 18 *Verilog-A/MS model for a relay.*

```
// Relay
`include "disciplines.vams"

module relay (p, n, ps, ns);
    parameter real thresh=0;                   // threshold (V)
    parameter real ron=0 from [0:inf);         // on resistance (Ohms)
    parameter real goff=0 from [0:/1ron);      // off conductance (Siemens)
    input ps, ns;
    electrical p, n, ps, ns;

    analog begin
        @(cross( V(ps,ns) – thresh, 0 ))
            ;
        if (V(ps,ns) > thresh)
            V(p,n) <+ ron*I(p,n);
        else
            I(p,n) <+ goff*V(p,n);
    end
endmodule
```

LISTING 19 *Verilog-A/MS model for a non-ideal relay.*

```
// Relay
`include "disciplines.vams"

module relay (p, n, ps, ns);
    parameter real thresh=0;                   // threshold (V)
    parameter real ron=0 from (0:inf);         // on resistance (Ohms)
    parameter real goff=0 from [0:/1ron);      // off conductance (Siemens)
    input ps, ns;
    electrical p, n, ps, ns;

    analog begin
        @(cross( V(ps,ns) – thresh, 0 ))
            $discontinuity(0);
        if (V(ps,ns) > thresh)
            I(p,n) <+ V(p,n)/ron;
        else
            I(p,n) <+ goff*V(p,n);
    end
endmodule
```

The discontinuity function could have also been employed in Listing 17 and Listing 18, but they are not needed in these cases because the discontinuity is caused by a switch branch. Any change in the overall state of a switch branch is assumed by the simulator to produce an order 0 discontinuity (regardless of whether it does or not).

6.2 Ideal Mechanical Stop

Another use for the switch branch is to model ideal barriers such as ideal diodes, or the mechanical equivalent, an ideal stop. The model for an ideal stop is given in Listing 20.

LISTING 20 *Verilog-A/MS model for an ideal mechanical stop.*

```
// Ideal mechanical stop
`include "disciplines.vams"

module barrier (p, n);
    inout p, n;
    kinematic p, n;

    analog begin
        @(cross((Pos(p,n) + F(p,n)), 0))
            ;
        if ((Pos(p,n) + F(p,n)) > 0)
            Pos(p,n) <+ 0;
        else
            F(p,n) <+ 0;
    end
endmodule
```

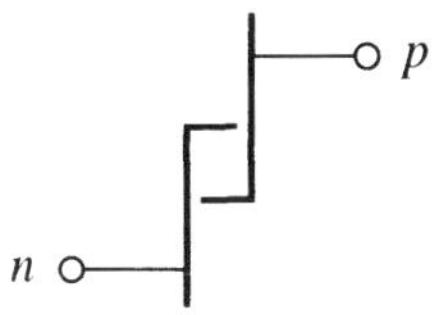

An ideal stop is a component that allows unrestricted motion up to a point, but does not allow motion beyond that point. Assume that point has a position of 0. Then the core of the stop model is

```
if (cond)
    Pos(p,n) <+ 0;
else
    F(p,n) <+ 0;
```

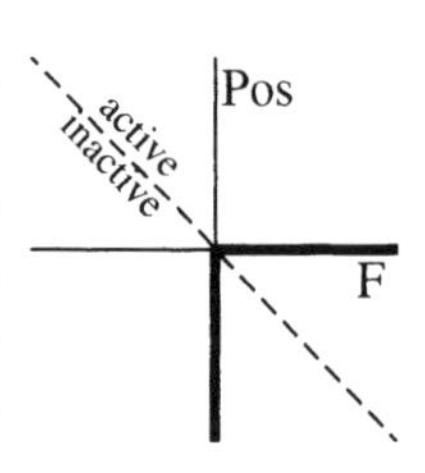

The modeling problem then becomes, what does one use for the condition. Using just the position or force alone results in numerical difficulties. This becomes obvious if one looks at the position versus force curve for the stop. Deciding whether the stop is active or not based solely on position is equivalent to dividing this plane in half, with the stop being active when in the right half of the plane. However, the curve never really leaves the right half of the plane; at best it follows its border. Given the tolerances associated with the simulation of nonlinear systems, such an approach is not robust. So instead, in this model the plane is cut on a diagonal as shown where the stop is active above the diagonal. Thus the condition becomes Pos(p,n) + F(p,n)) > 0. While this appears a bit strange because it combines position and force, it is more robust than using either position or force alone. The core of the model becomes

```
if (Pos(p,n) + F(p,n)) > 0)
    Pos(p,n) <+ 0;
else
    F(p,n) <+ 0;
```

To accurately resolve the points in time where the stop transitions between the active and inactive regions, a *cross* event is added as in the relay.

```
@(cross((F(p,n) + Pos(p,n)), 0))
    ;
```

Notice that the event statement has no action associated with it. It is present simply because the act of detecting the event will cause the simulator to place a time point very near the event, which will result in the transition being carefully resolved.

One could further improve this model by scaling the signals to make them consistently sized. To do so, one uses hierarchical names (5§9.4p230) to access the absolute tolerances associated with the signals (5§3.1.2p168), and the *abstol*s are used to scale the force so that it is the same relative size as the position. Then the *abstol* for position is passed into the *cross* function to be used as its tolerance (5§6.8.3p206).

```
real sum;

analog begin
    sum = Pos(p,n) + F(p,n)*(p.potential.abstol/p.flow.abstol);

    @(cross(sum, 0, , p.potential.abstol))
        ;
```

```
        if (sum > 0)
            Pos(p,n) <+ 0;
        else
            F(p,n) <+ 0;
    end
```

6.3 Ideal Diode

As an exercise, write a model for an ideal diode. When finished, compare what you wrote to Listing 21

LISTING 21 *Verilog-A/MS model for an ideal diode.*

```
// Ideal diode
`include "disciplines.vams"

module diode (a, c);
    inout a, c;
    electrical a, c;

    analog begin
        @(cross((V(a,c) + I(a,c)), 0))
            ;
        if ((V(a,c) + I(a,c)) > 0)
            V(a,c) <+ 0;
        else
            I(a,c) <+ 0;
    end
endmodule
```

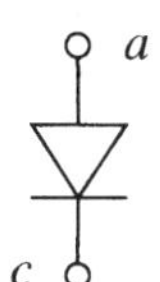

7 Voltage-Controlled Oscillator

A voltage-controlled oscillator (VCO) produces an output signal whose frequency is proportional to an input signal (generally a voltage, but it could also be a current),

$$f_{out} = Kv_{in}, \tag{29}$$

where K is the VCO gain given in units of Hertz per volt (or amp) and is often referred to as K_{vco}. The best way to model a VCO in a continuous-time setting is to integrate the input signal to compute the phase of the output signal,

$$\phi(t) = 2\pi\int Kv_{in}(t)dt, \tag{30}$$

and then produce the output signal from the phase. In this model, the output will be a sinusoid that is produced by passing the phase through the sine function,

$$v_{out}(t) = \sin\phi(t). \tag{31}$$

The Verilog-A/MS model that implements these equations is given in Listing 22. A new aspect of this model is that the input and output ports are declared using the voltage discipline.

LISTING 22 *Verilog-A/MS model for a sinusoidal VCO.*

```
// Voltage-controlled oscillator
`include "disciplines.vams"
`include "constants.vams"

module vco (out, in);
    parameter real Vmin=0;                              // minimum input voltage (V)
    parameter real Vmax=Vmin+1 from (Vmin:inf);         // maximum input voltage (V)
    parameter real Fmin=1 from (0:inf);                 // minimum output freq (Hz)
    parameter real Fmax=2*Fmin from (Fmin:inf);         // maximum output freq (Hz)
    parameter real ampl=1;                              // output amplitude (V)
    input in; output out;
    voltage out, in;
    real freq, phase;

    analog begin
        // compute the freq from the input voltage
        freq = (V(in) – Vmin)*(Fmax – Fmin) / (Vmax – Vmin) + Fmin;

        // bound the frequency (this is optional)
        if (freq > Fmax) freq = Fmax;
        if (freq < Fmin) freq = Fmin;

        // phase is the integral of the freq modulo 2π
        phase = 2*`M_PI*idtmod(freq, 0.0, 1.0, –0.5);

        // generate the output
        V(out) <+ sin(phase);

        // bound the time step
        $bound_step(0.1/ freq);
    end
endmodule
```

```
voltage out, in;
```

The voltage discipline is defined in *disciplines.vams* (5§2.4p159) to be

```
discipline voltage
    potential Voltage;
enddiscipline
```

Notice that this discipline is declared without a flow, making it a ***signal-flow*** discipline (as opposed to a conservative discipline), and the VCO a signal-flow model. Signal-flow models are generally used for more abstract models such as this. It is more natural to use the *voltage* discipline than the *electrical* discipline because in this model there is no mention of current (*electrical*'s flow). Signal-flow models can be freely connected to conservative models, and so there is no reason not to simplify a model by using signal-flow disciplines if possible.

Also notice that the output is listed before the input on the terminal list. It is a general Verilog (and SPICE) convention to place the outputs on the port list before the inputs. While it is not necessary to follow this convention, doing so will make your models a bit easier to use.

The parameter declarations are similar to those presented earlier, except that the defaults and limits for some of the parameters are defined in terms of previously defined parameters (5§2.3p157).

```
parameter real Vmin=0;
parameter real Vmax=Vmin+1 from (Vmin:inf);
parameter real Fmin=1 from (0:inf);
parameter real Fmax=2*Fmin from (Fmin:inf);
```

The default value and lower bound for *Vmax* is defined in terms of the value specified for *Vmin*. Thus if *Vmin* is given to be –0.5, then the default value for *Vmax* would be 0.5 and its lower bound would be –0.5. Similarly, the default value and lower bound for *Fmax* is defined in terms of the value specified for *Fmin*. These parameters are used to calculate k_{vco} and to bound the minimum and maximum output frequency.

The file *constants.vams* supplies `M_PI (5§2.1p152), which is replaced by the value of π and is used in

```
phase = 2*`M_PI*idtmod(freq, 0.0, 1.0, –0.5);
```

Verilog-A/MS provides the special function, ***idtmod***, that is used when modeling VCOs. It combines the operation of integration with respect to time with a modulus operation, and is sometimes referred to as a ***circular integrator***. The *idtmod* function takes five arguments of which all but the first are optional (5§4.6.3p179),

itdmod(*integrand*, *initial condition*, *modulus*, *offset*, *tolerance*)

and returns

$$y(t) = \underset{m}{\operatorname{mod}}\left(\left(\int_0^t x(\tau)d\tau + y_0 - b\right)\right) + b\,, \tag{32}$$

where y is the return value, x is the integrand, y_0 is the initial condition, m is the modulus, and b is the offset. In other words, the output value $y(t)$ satisfies

$$y(t) = \int_0^t x(\tau)d\tau + y_0 - k(t)m\,, \tag{33}$$

where $k(t)$ is an integer that is chosen so that $b \le y(t) \le b + m$, as shown in Figure 6. In this situation this function is preferred over *idt*, the conventional integration operator, because the output remains bounded, keeping the subsequent *sin* function in its accurate range. Even if a sinusoidal output is not desired, it is generally easier to produce the desired output wave shape with the bounded output of *idtmod* than if *idt* were used. Besides keeping its output bounded, the *idtmod* function is implemented in such a way that its internal state variable is also bounded, and so avoids tolerance and round-off problems.†

FIGURE 6 *Output of the idtmod operator when input argument is a constant α, the initial condition is y_0, the modulus is m, the offset is b, and k is an integer.*

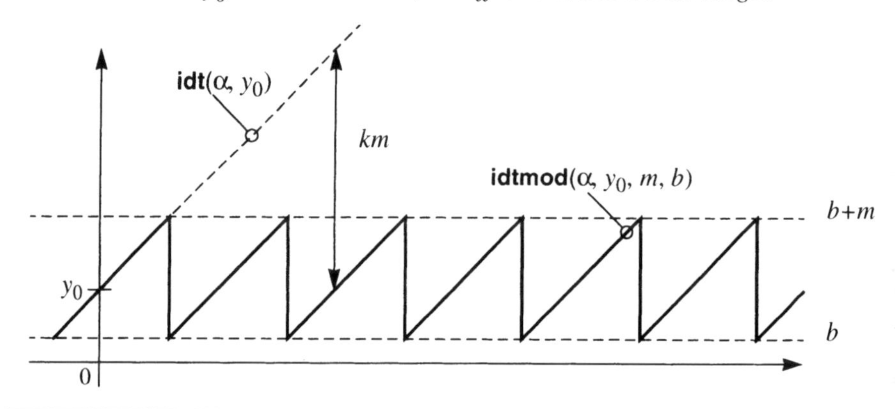

† An additional benefit of using the *idtmod* operator in the VCO is that it avoids a compatibility problem with RF simulators such as SpectreRF. When the input is a constant, the *idtmod* signals (both its internal state and its output) are periodic, whereas the signals associated with the *idt* operator are linear ramps. The RF simulator requires periodic solutions, which the *idtmod* operator provides and the *idt* operator does not.

One last new feature of Verilog-A/MS that is being used in this model is the *$bound_step* system function (5§5.1.1p190). It is a simulator directive, meaning that it affects the behavior of the simulator and not the behavior of the model. It simply directs the simulator to take time steps no larger than its argument. Be careful to assure the argument is not needlessly small, which would make the simulator run more slowly than necessary. Controlling the time step is required by this model because it produces changes in its output when there is no corresponding change in its input. As such, it has the responsibility to coordinate with the simulator to assure that the simulator does not simply miss the changes. For example, consider a model that produces a 1 GHz sine wave when the simulator has chosen to take 1 ns time steps. In this case the sine wave aliases to DC as a result of subsampling, and so the simulator would not actually notice that the sine wave was present, and so would not know to shrink its time step in order to follow it. Changes in the output generally take two forms, either there is an abrupt change or the change is smooth and continuous. Abrupt changes are generally triggered by the *timer* function (5§6.8.2p206), and so Verilog-A/MS is usually aware of when the change occurs and notifies the simulator so that it can choose its evaluation points accordingly. However, in the case of smooth changes, it is unable to provide the simulator any guidance and so it is up to the author of the model to explicitly provide the information the simulator needs, and that is done using the *$bound_step* function. Generally, it is necessary to use *$bound_step* when modeling autonomous systems (oscillators) with smoothly varying outputs, which of course is a perfect description of this VCO. Here, the time step is constrained to be no larger than one tenth the length of a period of the VCO's output.

8 Periodic Sample and Hold

A periodic sample and hold samples its input at evenly spaced points and produces an output that equals the value of the input at the most recent sample point.

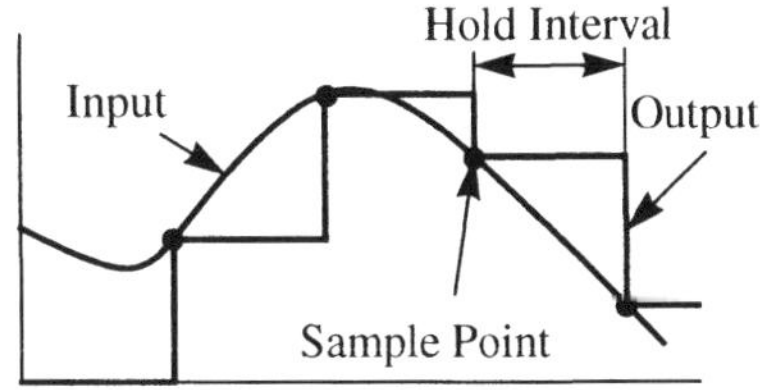

$$v_{\text{out}}(t) = v_{\text{in}}(nT) \quad (34)$$

where T is the sample period and n is the largest integer that is smaller than t/T, ($n = \lfloor t/T \rfloor$). An implementation is shown in Listing 23.

The act of sampling is an example of an event. It is an action that occurs instantaneously (it has no duration). At least, that is the way it will be modeled. In this case, the sampling is periodic, and so represents a stream of spontaneous events. This activity is realized with

LISTING 23 *Verilog-A/MS model for an ideal periodic (self-clocked) sample and hold.*

```
// Self-clocked (periodic) sample and hold
`include "disciplines.vams"

module sh (p, n, ps, ns);
    parameter real period=1 from (0:inf);       // sampling period (s)
    parameter real toff=0 from [0:inf);         // offset time for sampling (s)
    output p, n; voltage p, n;                  // output port
    input ps, ns; voltage ps, ns;               // input port
    real save;

    analog begin
        // Sample the input
        @(timer(toff, period) or initial_step) begin
            save = V(ps,ns);
            $discontinuity(0);
        end

        // Produce the output
        V(p,n) <+ save;
    end
endmodule
```

```
@(timer(toff, period) or initial_step)
    save = V(ps,ns);
```

This event statement is triggered by events produced by either the *timer* function (5§6.8.2p206) or by *initial_step* (5§6.8.1p205). The *timer* function produces an event at the (absolute) time given by its first argument, and if its second argument is present, it will produce subsequent events with a period equal to the value of the second argument. In other words, events are produced at $t_{off} + nT$, where t_{off} is the value of the first argument, T is the value of the second, and $n = 0, 1, 2, \ldots$. The second source of events is *initial_step*, a built-in event source that produces a single event at the beginning of the simulation time interval. It is generally used for initialization, and this case initializes the sample and hold. The events produced by these event sources are combined using the *or* keyword. When an event occurs, the event statement executes its event clause, in this case a statement that assigns the value of the input to a local variable, *save*.

It is important to recognize that the value of the *save* variable is set only when events occur, but its value is used at all points in time. As such, the variable must be able to remember its value. All Verilog-A/MS variables have this ability, but not all use it.

Those that retain their value from a previous point in time are referred to as ***state variables***.

The event clause also contains a simulator directive, *discontinuity* (5§5.1.2p191). Like *$bound_step*, its presence indicates that something is occurring that the simulator should be aware of this. In this case, the *discontinuity* statement announcing that the output waveform produced by the module contains a discontinuity. The argument indicates the order of the discontinuity. In this case, the argument is 0, indicating that the output waveform itself is discontinuous (you could not trace the waveform with your finger without raising your finger and moving it at the discontinuity). If the value of the argument is k with $k > 0$, it indicates that the discontinuity occurred in the kth derivative. Thus, if $k = 1$, the discontinuity occurs in the slope of the waveform (it has a noticeable kink, but does not jump). The simulator uses this information to optimize the operation of its internal algorithms.

8.1 Smoothing the Output

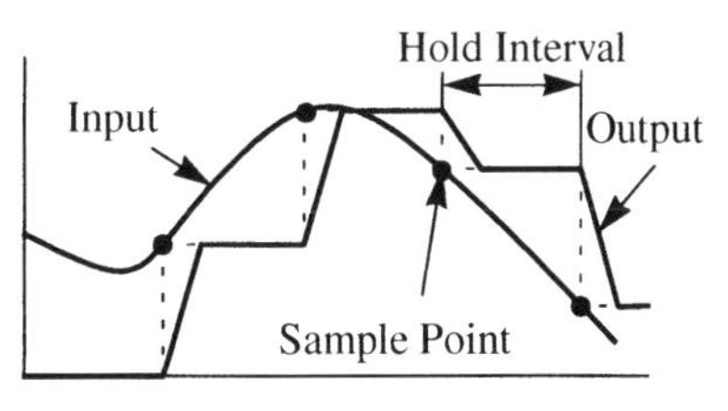

Rather than producing models whose outputs jump discontinuously, it is often prudent to smooth the outputs to eliminate the abrupt jumps. This is particularly true if the output of the module could drive a dynamic component like a capacitor or inductor, which often behaves badly when faced with abruptly discontinuous waveforms. It is for this reason that the sample and hold module was rewritten so as to eliminate the jumps, as shown in Listing 24.

The magic is performed by a new function, *transition*, which is often referred to as the *transition filter* (5§4.6.4p180).

```
V(p,n) <+ transition(save, td, tt);
```

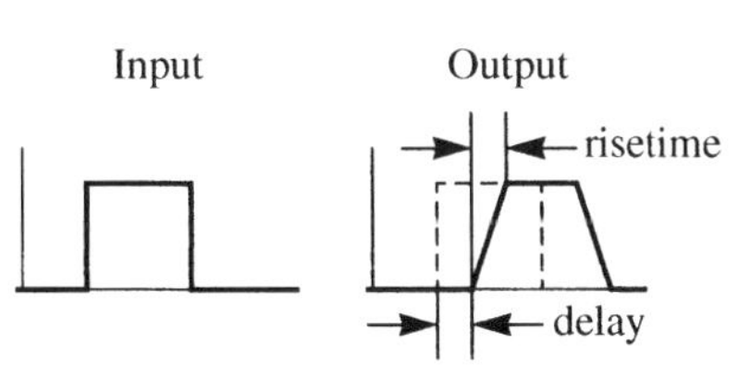

The *transition* function takes four arguments, the first being a piecewise constant waveform. It is this waveform that will be smoothed and transferred to the output. Be sure that this first argument is constant valued except at isolated points where it jumps abruptly. The *transition filter* does not react well when this argument varies smoothly as a function of time. The remaining arguments are optional. The second is the delay or the time it takes for the transition to pass through the filter. The third and fourth arguments specify the transition time for the transition on the output, the third being the rise time (used when the output is becoming more positive) and the fourth being the fall time. If the

LISTING 24 *Verilog-A/MS model for a periodic (self-clocked) sample and hold.*

```
// Self-clocked (periodic) sample and hold
`include "disciplines.vams"

module sh (p, n, ps, ns);
    parameter real period=1 from (0:inf);        // sampling period (s)
    parameter real toff=0 from [0:inf);           // offset time for sampling (s)
    parameter real td=0 from [0:inf);             // delay from sampling to output (s)
    parameter real tt=period/100 from [0:inf);    // duration of output transitions (s)
    input ps, ns; voltage ps, ns;                 // input port
    output p, n; voltage p, n;                    // output port
    real save;

    analog begin
        // Sample the input
        @(timer(toff, period) or initial_step)
            save = V(ps,ns);

        // Produce output with well-controlled transitions
        V(p,n) <+ transition(save, td, tt);
    end
endmodule
```

fourth is missing, the fall time is taken to be the same as the rise time. If the third is missing or zero, it is taken to be the value of the `default_transition compiler directive.

Even though the first derivative of the output waveform is discontinuous, this module does not need a *discontinuity* statement as the last one did because the *transition* function takes responsibility for notifying the simulator of the discontinuities that it produces. In addition, it directs the simulator to place time points at the corners of the transition (in this case, at *td* and *td+tt*). However, the transition filter behaves a bit differently if the transition time, tt, is specified to be zero. In this case the transition time actually used by the filter is the one specified in the currently active `default_transition compiler directive, but no attempt is made to resolve the trailing corner of the transition.

9 Time Interval Measurement

When constructing test benches it is often useful to be able to accurately measure the timing of events such as delays. An example of that is given here in the form of a time interval measurement. It monitors two signals, *start* and *stop*, and measures the time between when *start* and *stop* cross a given threshold,

$$\delta = t_1 - t_0, \tag{35}$$

where t_0 and t_1 are the time at which *start* and *stop* cross the threshold. An estimate of the mean time interval is computed using,

$$\mu_\delta \cong \frac{\sum_{n=1}^{N} \delta_n}{N}. \tag{36}$$

The time interval measurement is implemented in Listing 25. The first aspect of this model that is worthy of comment is the declaration for the *dir* parameter.

```
parameter integer dir=1 from [–1:1] exclude 0;  // dir=1 for rising edges
                                                // dir=–1 for falling edges
```

The *dir* parameter is used to specify whether the module should time rising or falling edges of the signals on *start* and *stop*. It constrains the valid choices to only –1 and +1 by setting the parameter type to integer, setting the range to include all integers between –1 and +1, and then finally excluding 0 from the range (5§2.3p157).

This model has two modes, unarmed and armed. It initially starts unarmed and is armed when a valid threshold crossing is detected on the *start* input. The time of that threshold crossing is measured and stored in t0. That task could be accomplished by the following code,

```
t0 = last_crossing(V(start) – thresh, dir);
```

The *last_crossing* function monitors its first argument and returns the time that it last crossed 0 in the specified direction (5§4.7.2p188). The simulator executing the model only evaluates the model at discrete points in time, and so *last_crossing* uses interpolation to estimate when the crossing occurred. If used on its own, it is not likely to be very accurate.

Unlike *last_crossing*, *cross* coordinates with the simulator to assure that the simulator places an evaluation point very near the threshold crossing. So one could use the following code in place of the code given above,

```
@(cross(V(start) – thresh, dir)) begin
    armed = 1;
    t0 = $abstime;
end
```

LISTING 25 *Verilog-A/MS model that measures and saves the time interval between transitions on two signals.*

```
`include "disciplines.vams"

module time_interval_measurement (start, stop);
    parameter real thresh=0;  // threshold (V)
    parameter integer dir = 1 from [-1:1] exclude 0;
                                    // 1 for rising edges, -1 for falling
    input start, stop;
    voltage start, stop;
    integer count, armed;
    real t0, t1, sum, mean;

    analog begin
        // sense and record the start of the interval
        t0 = last_crossing(V(start) - thresh, dir);
        @(cross(V(start) - thresh, dir))
            armed = 1;

        // sense and record the end of the interval
        t1 = last_crossing(V(stop) - thresh, dir);
        @(cross(V(stop) - thresh, dir)) begin
            if (armed) begin
                armed = 0;
                count = count +1;
                sum = sum + (t1 - t0);
            end
        end

        // produce the final report at end of simulation
        @(final_step) begin
            $strobe("time interval measurements = %d.\n", count);
            if (count) begin
                mean = sum / count;
                $strobe("mean time interval (est)= %g.\n", mean);
            end else
                $strobe("Could not measure time interval.\n");
        end
    end
endmodule
```

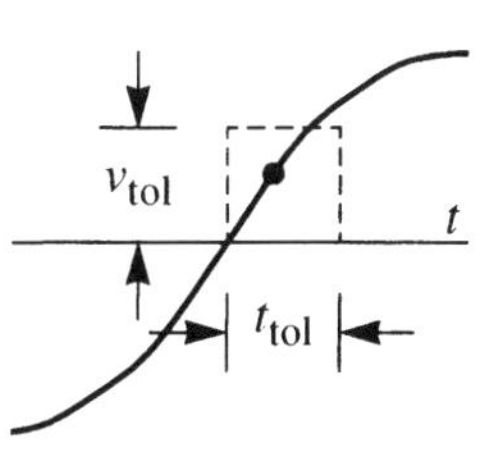

In this case the time of the crossing is recorded by evaluating *$abstime* in the event clause (*$abstime* evaluates to the current value of time as a real number (5§4.5.1p175)). While this would be more accurate than using *last_crossing* alone, its accuracy is not as good as it could be because the simulator places the point where the event clause is evaluated slightly after the threshold crossing to assure that the crossing actually occurs. The evaluation point is guaranteed to fall within a box whose size is specified using optional tolerance parameters. t_{tol} and v_{tol} are the optional third and fourth parameters to the *cross* function. t_{tol} is an absolute tolerance given in terms of time, and v_{tol} is the same except given in terms of the value of the signal (the first argument). It is possible to increase the accuracy of this approach by specifying the tolerances. Generally specifying t_{tol} is sufficient; however there are limits to how tight the tolerances can be before they become impossible for the simulator to satisfy.

For the best accuracy and efficiency, it is best to combine the two approaches given above,

```
t0 = last_crossing(V(start) – thresh, dir);
@(cross(V(start) – thresh, dir))
    armed = 1;
```

In this way the *last_crossing* function benefits from the *cross* function causing the simulator to place an evaluation point very near the threshold crossing. Together, they are considerably more accurate that either apart. And if the accuracy of the above is not sufficient, one can tighten the tolerances on the *cross* function.

One might wonder why the *last_crossing* function is not placed within the event clause of the *@cross* statement. After all, its return value is only used in within the clause, and doing so would seem to make the model more efficient as the *last_crossing* function would only be evaluated at the threshold crossings and not on every simulator evaluation point. However it is important to realize that *last_crossing* is not a simple scalar function like *sqrt* or *sin*. It must monitor its input over time in order to determine that a crossing has occurred and precisely when it occurs. If it were contained in the event clause, it would only be able to observe its inputs at isolated points in time and so would not know when the crossings occurred, or even if they had occurred. As such, there are constraints on where a *last_crossing* function can be used, and it is not the only function or operator that is similarly constrained (5§4.6p177). These restrictions are discussed further in Section 10.

This model collects data through out the simulation interval; and at the end analyzes the data and writes it to the screen. It uses a built-in event named *final_step* to properly identify the end of the simulation interval (5§6.8.1p205). Within the event clause,

$strobe is used to write the results to the screen. The *strobe* function is very much like the *printf* function in the C programming language (5§5.2.1p192). The first argument is a string, which is printed after the percent codes are replaced by the subsequent arguments. For example,

```
$strobe("delay measurements = %d.\n", count);
```

prints the string "delay measurements = 100" where '%d' was replaced by the value of count, here assumed to be 100. The *d* in '%d' indicates that the argument should be formatted using a decimal notation. The '\n' is an escape sequence that ends the line (5§2.1p152).

10 Analog to Digital Converter

An analog-to-digital converter (ADC) is a component that takes a continuous-value signal at its input and converts it to an integer after appropriate scaling. The integer is produced in the form of a binary number on a bus. There are many different types of ADCs, each with different characteristics. A simple Nyquist converter is given in Listing 26. It is a clocked converter that produces an output on each clock edge without latency.

This model uses a ***bus***, or ***vector port***, to output its result (5§2.5p164). The bus is declared with

```
output [0:bits–1] out;
voltage [0:bits–1] out;
```

The first line declares the direction of the port, which is required of all ports. The second associates the *voltage* discipline with the port, which is needed because the port is accessed from within the behavioral part of the model. The port is declared as a bus by giving the limits of the indices for the individual members of the array. In this case, the bus consists of members *out*[0], *out*[1], ..., *out*[*bits*–1]. Since *bits* is used in the declaration of the out, it must be declared first. This module also declares an integer vector,

```
integer result[0:bits–1];
```

As with the port, this defines an array that consists of the members *result*[0], *result*[1], ..., *result*[*bits*–1]. This module also declares a new type of variable,

```
genvar i;
```

A ***genvar*** is a restricted integer variable that is used as an index in *for* loops (5§2.2p155). It is not necessary to use a *genvar* as an index to a *for* loop, but doing so acts to constrain the behavior of the loop, making it possible to include operators in

LISTING 26 *Verilog-A/MS model for an N-bit analog-to-digital converter.*

```
// N-bit Analog to Digital Converter
`include "disciplines.vams"

module adc (out, in, clk);
    parameter integer bits = 8 from [1:24]; // resolution (bits)
    parameter real fullscale = 1.0;          // input range is from 0 to fullscale (V)
    parameter real td = 0;                   // delay from clock edge to output (s)
    parameter real tt = 0;                   // transition time of output (s)
    parameter real vdd = 5.0;                // voltage level of logic 1 (V)
    parameter real thresh = vdd/2;           // logic threshold level (V)
    parameter integer dir = 1 from [-1:1] exclude 0;
                                             // 1 for rising edges, -1 for falling
    input in, clk;
    output [0:bits-1] out;
    voltage in, clk;
    voltage [0:bits-1] out;
    real sample, midpoint;
    integer result[0:bits-1];
    genvar i;

    analog begin
        @(cross(V(clk)-thresh, +1) or initial_step) begin
            sample = V(in);
            midpoint = fullscale/2.0;
            for (i = bits - 1; i >= 0; i = i - 1) begin
                if (sample > midpoint) begin
                    result[i] = vdd;
                    sample = sample - midpoint;
                end else begin
                    result[i] = 0.0;
                end
                sample = 2.0*sample;
            end
        end

        for (i = 0; i < bits; i = i + 1) begin
            V(out[i]) <+ transition(result[i], td, tt);
        end
    end
endmodule
```

the loop that would otherwise find the loop inhospitable. In particular, a *genvar* variable can only be assigned within the *for* loop control section (not within the body of the *for* loop). In addition, assignments to a *genvar* variable can consist only of expressions of static values: the expressions can consist of operations on parameters, literals, and other *genvar* variables. These restrictions result in the bounds of the loop being static. As such, they are known before the simulation begins and they cannot change during the simulation. Conceptually, this allows the loop to be unrolled in advance.

The *genvar* variable *i* is used as the index in the loop

```
for (i = 0; i < bits; i = i + 1) begin
    V(out[i]) <+ transition(result[i], td, tt);
end
```

Notice that the loop iterates through $i = 0, 1, \ldots, bits - 1$, where *bits* is a constant. Thus, this loop is equivalent to

```
V(out[0]) <+ transition(result[0], td, tt);
V(out[1]) <+ transition(result[1], td, tt);
   :            :             :
V(out[bits–1]) <+ transition(result[bits–1], td, tt);
```

In this way, each iterate of the loop gets its own *transition* filter. The *transition* filter is an analog operator, meaning that it maintains a history of its argument so that the output of the filter can depend on past values of its input. As such, it is important that there be one transition filter for each signal, or in other words, one *transition* filter for each iterate of the loop. Without the restrictions imposed on the loop by the *genvar* index, it would not be possible to statically associate the filters with their inputs and outputs. For reasons that are more difficult to explain, from within analog processes any access to the analog signals in busses also must be statically associated, and so can only be included in loops that are restricted by having a *genvar* index. This is true regardless of whether the signals are being observed or driven.

The remainder of the model is enclosed in an event clause that is triggered by the clock edge. When triggered, it samples the input and then enters a loop that sequentially determines the value of each bit, from most significant to least. The bit is set high if the sample is above the midpoint. In this case, the sample is reset to be the difference between its original value and the midpoint. Otherwise the bit is set low. In either case, the value of the sample is doubled and the procedure repeats to determine the value of the next most-significant bit, and the process continues until all of the values of the bit values are known and stored in *results*. All of this occurs as the clock signal passes through its threshold. At all other times, the bit values stored in the *results* vector are passed to the *out* vector through the *transition* filters, which are responsible for adding the desired delay and smoothing the output transitions.

11 Digital to Analog Converter

A digital-to-analog converter (DAC) performs the inverse of the operation performed by an ADC; it converts an N-bit binary integer into a real-valued signal. An implementation is given in Listing 27.

LISTING 27 *Verilog-A/MS model for an N-bit digital-to-analog converter.*

```
// N-bit Digital to Analog Converter
`include "disciplines.vams"

module dac (out, in, clk);
    parameter integer bits = 8 from [1:24]; // resolution (bits)
    parameter real fullscale = 1.0;         // output range is from 0 to fullscale (V)
    parameter real td = 0;                  // delay from clock edge to output (s)
    parameter real tt = 0;                  // transition time of output (s)
    parameter real vdd = 5.0;               // voltage level of logic 1 (V)
    parameter real thresh = vdd/2;          // logic threshold level (V)
    parameter integer dir = 1 from [–1:1] exclude 0;
                                            // 1 for rising edges, –1 for falling
    output out;
    input clk;
    input [0:bits–1] in;
    voltage out, clk;
    voltage [0:bits–1] in;
    real aout;
    integer weight;
    genvar i;

    analog begin
        @(cross(V(clk) – thresh, dir) or initial_step) begin
            aout = 0;
            weight = 2;
            for (i = bits – 1; i >= 0; i = i – 1) begin
                if (V(in[i]) > thresh) begin
                    aout = aout + fullscale/weight;
                end
                weight = weight*2;
            end
        end
        V(out) <+ transition(aout, td, tt);
    end
endmodule
```

This introduces nothing that is conceptually new. Like the ADC, it monitors the clock and triggers an event clause when the clock passes through the threshold in the specified direction. At this time, an algorithm is run that scans through the bits, determines their weight, and accumulates that weight to the eventual output, *aout*, if the bit is set high. At all other times, *aout* is passed to the output through a *transition* filter, which adds any needed delay and smoothes the transitions. The *for* loop in the event clause contains an indexed access to the analog signals on a bus, and so uses a *genvar* index.

12 Lossy Inductor

A model for an inductor that is accurate over a broad range of frequencies is shown in Figure 7. Component H in the model represents the skin-effect loss. The impedance of the skin effect is given by,

$$Z(f) = \frac{\sqrt{jf}}{H}. \tag{37}$$

This model involves a non-integer power for f and so is a distributed model. In other words, $Z(f)$ cannot be exactly represented using a finite number of poles and zeros and the model cannot be implemented exactly by combining a finite number of lumped components (resistors, capacitors, and inductors).

FIGURE 7 *RF inductor model.*

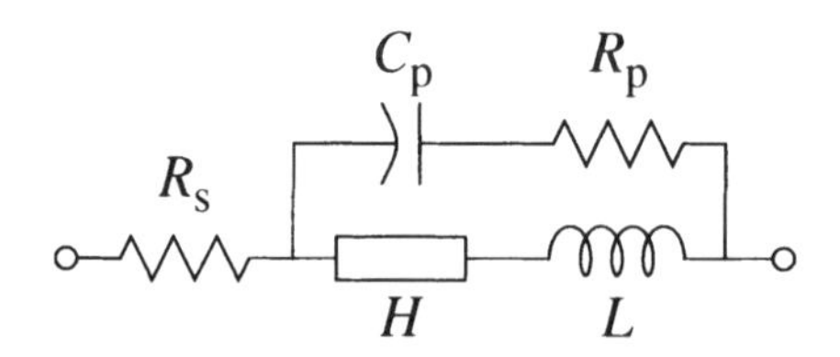

An impedance of $\sqrt{jf}$ is approximated over a finite range of frequencies with an equal number of real poles and zeros alternating and evenly spaced in a logarithmic sense over that range, as shown in Figure 8. The range of the approximation is from f_0 to f_1, with the impedance flattening out at frequencies outside of this range. The range of frequencies over which skin effect must be accurately modeled can be determined by examining its contribution to the overall impedance of the inductor. The impedance of a representative inductor is shown in Figure 9. Notice that the impedance is separated into its real and imaginary parts (the resistance and the reactance). The resistance represents the loss. It is important to accurately model the resistive portion in the operating frequency range even where the reactive portion is much larger

FIGURE 8 *Modeling the $\sqrt{jf}$ nature of skin effect using a collection of real poles and zeros.*

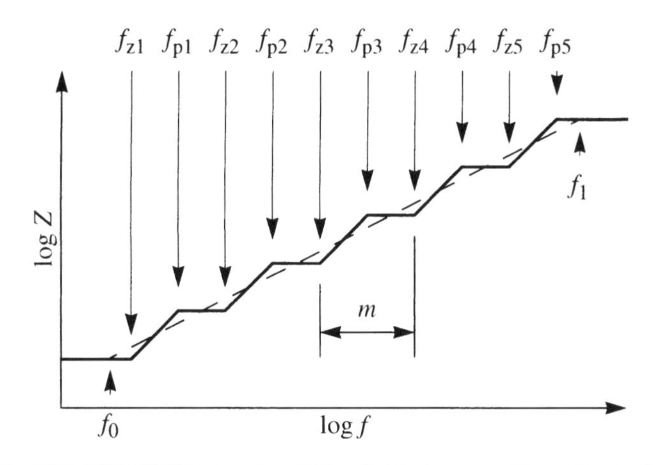

FIGURE 9 *Modeled impedance of an RF inductor separated into real (R) and imaginary (X) parts.*

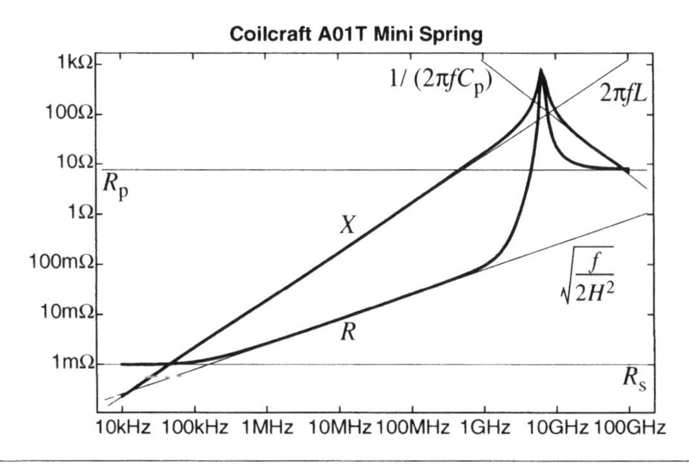

because it determines the Q of resonators and because designers sometimes use tuning to cancel out the reactive portion of the inductor.

Because the operating frequency range is not known a priori, we must model the skin effect over the range of frequencies where it is significant. The low frequency bound, f_0, is chosen to be the frequency where the impedance of the skin effect first dominates over R_s. Since our lumped approximation of skin effect naturally flattens out at low frequency, R_s will be combined with H to form H_R as shown in Figure 10. Notice

FIGURE 10 *Modified RF inductor model.*

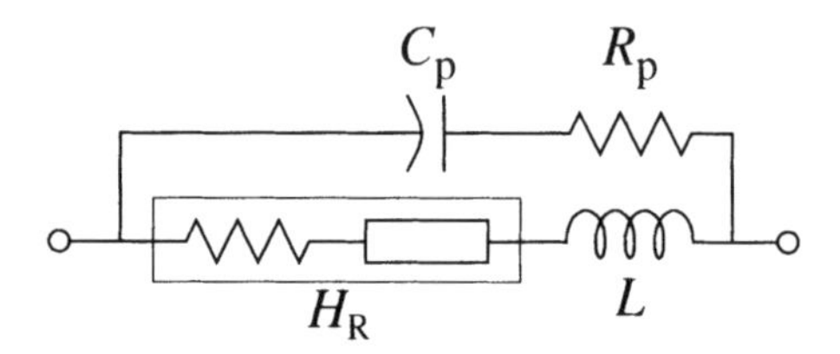

that the structure of the model was modified slightly to accommodate the combining of R_s and H. Since R_p is always much larger than R_s, the difference is not expected to be significant.

The low frequency bound, f_0, is the frequency where the resistance (the real part) of the skin effect begins to dominate over R_s,

$$f_0 = 2(R_s H)^2 . \tag{38}$$

The high frequency bound, f_1, is chosen to be the resonant frequency of LC_p, because above this frequency the capacitive path through the inductor dominates over the inductive path, and so R_p dominates over H,

$$f_1 = \frac{1}{2\pi\sqrt{LC_p}} . \tag{39}$$

Placing f_1 at the resonant frequency L and C_p does act to increase the modeled Q of this parasitic resonance, but does not significantly affect the Q in the operating frequencies of the inductor.

The number of lumps, n, is chosen to provide a sufficient level of accuracy. The larger n, the closer the impedance of the lumped approximation matches $\sqrt{jf}$. Typically, using 1½ lumps per decade of frequency range gives a good fit (but 1 lump per decade is often sufficient),

$$n = \left\lceil \frac{3}{2}\log\frac{f_1}{f_0} \right\rceil . \tag{40}$$

The frequency span of one lump is then

$$m = \sqrt[n]{\frac{f_1}{f_0}}. \tag{41}$$

From Figure 8, the zero and pole frequencies are simply

$$f_{zi} = m^{i-\frac{3}{4}} f_0 \text{ and} \tag{42}$$

$$f_{pi} = m^{i-\frac{1}{4}} f_0 \tag{43}$$

with i ranging from 1 to n. For the inductor shown in Figure 9, impedance of the model for just the skin effect is shown in Figure 11. In this example, f_0 = 25 kHz and f_1 = 65 GHz. It is clear that accuracy degrades at the ends of the range. If accuracy at the ends is important, consider lowering f_0 and/or raising f_1 by and order of magnitude.

FIGURE 11 *Impedance of skin effect model.*

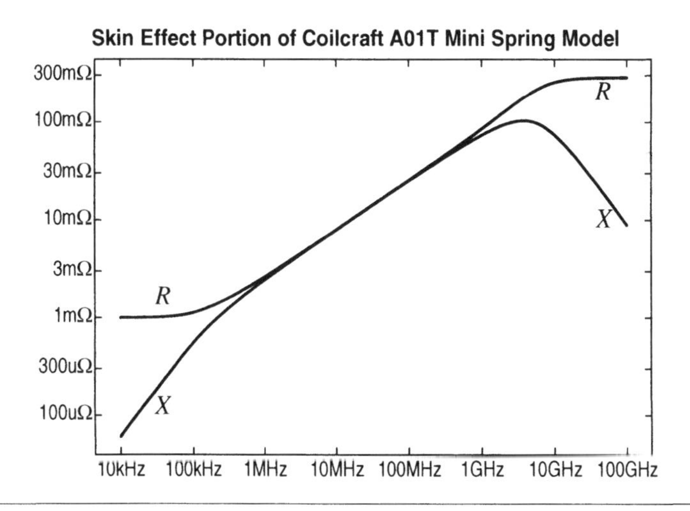

A Verilog-A/MS model of the inductor itself implements Figure 10 and is given in Listing 28. It uses (38) and (39) to compute f_0 and f_1 for the skin effect model. The parameter *esr* specifies R_s and *rcp* specifies R_p.

Listing 29 shows the module that implements the skin effect model. It is based on (40), (41), (42), and (43). The model starts by computing the location of the poles and

LISTING 28 *Verilog-A/MS model of the lossy inductor.*

```
`include "discipline.vams"
`include "constants.vams"

module lossy_ind (t1, t2);
    parameter real esr=0.01 from (0:inf);   // equivalent series resistance (Ohms)
    parameter real rcp=1.0 from (0:inf);    // series resistance of shunt C (Ohms)
    parameter real cp=10f from (0:inf);     // shunt capacitance (F)
    parameter real l=1n from (0:inf);       // inductance (H)
    parameter real h=70K from (1:inf);      // magnitude of skin effect at 1Hz (Ohms)
    electrical t1, t2;
    inout t1, t2;
    electrical n1, n2;

    resistor #(.r(rcp)) Rp (n2, t2);
    skin_effect #(.r0(esr), .f0(2*esr*esr*h*h), .f1(1/(2*`M_PI*sqrt(l*cp)))) Hr (t1, n1);
    capacitor #(.c(cp)) Cp (t1, n2);
    inductor #(.l(l)) L (n1, t2);
endmodule
```

zeros needed to achieve the transfer function shown in Figure 8, and then implements the transfer function using the *laplace_zp* function.

This model starts by declaring two real arrays. These arrays will hold the poles and zeros for the Laplace transform function that implements the module's transfer function. The array index bounds are specified by expressions, the terms of which are either constants or parameters; a requirement when declaring arrays.

The analog process begins with an event statement that is triggered on the initial step of any static analysis (5§6.8.1p205). In this case, *initial_step* takes an argument, "static", that indicates which analysis, or class of analyses, must be active for the event to be generated. In other words, in this case, the *initial_step* is only produced on the first step of a static analysis. The term static refers to any equilibrium point calculation, including a DC analysis as well as those that precede another analysis, such as the DC analysis that precedes an AC or noise analysis, or the initial condition analysis that precedes a transient analysis.

The *initial_step* event triggers the calculation of the poles and zeros used by the Laplace filter. Since the poles and zeros never change, they only need to be calculated once at the beginning of the analysis. As such, the *initial_step* event is used to increase the efficiency of the model. Without it, the poles and zeros would be recalculated at every time point, a substantial waste of time.

LISTING 29 *Verilog-A model of the skin effect.*

```
`include "discipline.vams"
`include "constants.vams"

module skin_effect (p, n);
    parameter integer lumps = 10 from (1:30];  // number of lumps in approximation
    parameter real f0=1 from (0:inf);          // lower frequency bound (Hz)
    parameter real f1=10 from (f0:inf);        // upper frequency bound (Hz)
    parameter real r0=1 from (0:inf);          // DC series resistance (Ohms)
    electrical p, n;
    inout t1, t2;
    real mult, mult2, wp, wz;
    real zeros[0:2*lumps-1], poles[0:2*lumps-1];
    integer i;

    analog begin
        @(initial_step) begin
            mult = pow(f1/f0, 1.0/(4*lumps));
            mult2 = mult*mult;
            wz = 2*`M_PI*mult*f0;
            wp = mult2*wz;
            for (i=0; i < lumps; i=i+1) begin
                zeros[2*i] = -wz;
                zeros[2*i+1] = 0;
                poles[2*i] = -wp;
                poles[2*i+1] = 0;
                wz = mult2 * wp;
                wp = mult2 * wz;
            end
        end
        V(p,n) <+ r0*laplace_zp( I(p,n), zeros, poles );
    end
endmodule
```

The event clause contains a *for* loop, an iterative statement (5§6.6.2p202). The statement

```
    for (i=0; i < lumps; i=i+1) begin
          ⋮
    end
```

is conceptually equivalent to

```
i = 0;
while (i < lumps) begin
        :
    i = i + 1;
end
```

meaning that when executing the *for* loop, the first expression is evaluated as an initializer, the second is used as a condition that causes the loop to continue as long as it remains true, and the advancement expression, which specifies how the successive iteration will differ from the current one. In this case, the variable *i* is acting as an iterator that progresses from 0 to lumps–1. Once *i* reaches a value that equals *lumps*, the condition is no longer satisfied and the loop terminates.

Upon termination of the event statement the *poles* and *zeros* arrays are initialized. They are passed to the *laplace_zp* operator that actually implements the behavior of the module (5§4.6.7p182). This operator takes a signal, in this case I(p,n), and filters that signal with a transfer function with the given poles and zeros to produce the output. The poles and zeros are given in arrays of pairs of real numbers. The first number in each pair is the real part of the critical frequency and the second is the imaginary part. For this model, the poles and zeros are all real, and so the second number is always zero. If any of the critical frequencies were complex, then they would have to be given as conjugate pairs. In other words, if there is a pole (or zero) at $s = \alpha + j\omega$ with $\omega \neq 0$, then there must also be another pole (or zero) specified at $\alpha - j\omega$

13 Tolerances

The behaviors included in the analog processes in a Verilog-A/MS description are combined into a nonlinear system of differential and algebraic equations (DAEs), which are passed to a simulator to be solved. These equations are solved numerically in a two step process. Generally, the first step is to discretize them in time. In other words, the differential operators are replaced by finite-difference approximations. This allows the continuous waveforms to be discretized in time with waveforms that have a finite number of time points. In doing so, the single system of nonlinear DAEs has been replaced by a sequence of nonlinear algebraic equations. These sets of nonlinear equations are then solved using an iterative approach, such as Newton's method. In these approaches, a guess at the solution is iteratively refined until it is sufficiently accurate. In both steps, tolerances are used to guide the process. When discretizing time the error increases with the size of the time step, as does the simulation speed. Tolerances are used to determine how small the steps must be made to achieve a certain level of accuracy. Similarly, when applying an iterative solver to the nonlinear equations, a criterion is needed for stopping the iteration once the desired level of accuracy is achieved.

Generally there are two tolerances that are needed, a relative tolerance and an absolute tolerance. Most people are familiar with the relative tolerance, *reltol* in SPICE. When one says they are expecting less than 1% error, they are using a relative tolerance, because the error in the desired quantity is specified relative to the value of the quantity itself. Generally, there is only one relative tolerance and it is used globally by the simulator. Verilog-A/MS provides no way to specify the relative tolerance. Instead, the simulator itself must provide some mechanism for allowing the user to specify the relative tolerance.

Relative tolerances work well as long as there is some representative value with which to compare the error. Sometimes that is not possible. For example, consider the case of applying a relative tolerance to a signal whose error-free value is zero. In this case, a relative tolerance will insist that the error also be zero, which is generally not possible to achieve. For this reason, simulators also use an absolute tolerance. The absolute tolerance gives the minimum interesting size of a quantity. By definition, anything smaller than the absolute tolerance is uninteresting and so can be ignored. The minimum interesting size of a quantity varies with the type of quantity. Typically in SPICE, any voltage smaller than 1 μV is considered negligible and so is ignored, whereas with current it is anything smaller than 1 pA that is negligible. It is for this reason that Verilog-AMS allows absolute tolerances to be specified as part of a *nature* where it is given in the *abstol* field.

In Verilog-AMS the issue of tolerances is mainly a matter of getting them from where they are known to where they are needed. The relative tolerance is known globally, so it is not an issue. The absolute tolerances are specified in the natures. The tolerances are needed in several places. They must be accessible to the nodes and branches so that the simulator can confirm convergence by assuring that Kirchhoff's laws are satisfied. They also needed by the differential operators (*ddt*, *idt*, *laplace*, etc.) where they control the discretization of time. All nodes and branches have *disciplines* that provide direct access to the absolute tolerance through their *natures*. For the differential operators, things are not so easy.

During the development of the language the idea of restricting the differential operators in such a way that they had direct access to the nodes and branches was considered, but it was felt that doing so would place too much burden on the model developers. And so, the burden now falls on the implementation of Verilog-AMS to try to propagate the tolerances to the differential operators. How this occurs is left unsaid. In some cases it might be possible for the operator to pick up the tolerances from its argument. In other, it might be able to access the tolerances at its output. It is to support these cases that *natures* have the ability to refer to other natures. The operators can use the *ddt_nature* and *idt_nature* fields to access tolerances that they might need. However, there are no guarantees that the operators will be able to gain access to the tolerances. In these cases, you have the option of specifying the tolerances explicitly on the operator. Each of the operators allows the tolerance to be specified as a number or as the name of a nature from which the tolerance is extracted.

As an alternative to specifying tolerances, one can determine what tolerances the simulator is using and simply scale signals to be compatible. This is generally not preferred because it makes the models harder to write and the results harder to interpret.

14 Elements of Style

Within the syntactic constraints of the language and the requirements of the model, Verilog-AMS, like all languages, gives you a great deal of freedom in determining the overall look and readability of your model. You are free to provide comments, arrange the white space, choose the identifier names and the order in which they are declared, and so forth. The way you make these choices constitutes your style. To some degree, style is a personal choice based on aesthetics. However, a style should also be consciously designed to make your models easily understandable; both to yourself and to others. The style used in this book makes a good starting point. It is designed to be both informative and to allow one to quickly see the structure of the model; and it is comprehensive enough to provide consistency to your models but not so verbose as to be burdensome.

Our style is illustrated in Listing 30. It starts with a comment header that describes the model. This header is missing from most of the models given in the book to save space, but it is an important part of any shared model. It should include comments about the model itself, plus information about the author, the version, and perhaps where to get more information (references to papers or web sites, etc.). Consistent placement and formatting of the sections of this header allows readers to quickly skip over sections that are not of interest.

The module itself necessarily begins with declarations. Often your are free to choose any order you like when declaring parameters, ports, and variables except when the declaration of an object depends on the value of a parameter. This could happen if the value of the parameter is used when specifying the size, default value, or bound for the object. In this case the parameter must be declared first. It is for this reason, and because parameters are of high interest to users of your model, that the parameters are declared first. When declaring parameters, you should include range limits unless the valid range is unbounded. Doing so makes your models more robust and aids in the

LISTING 30 *A varactor model that illustrates the modeling style used in this book.*

```
// Varactor Model
//
// Implements: c = c0 + c1* tanh((v – v0)/v1)
//
// Version 1a, 19 June 02
// Ken Kundert
//
// Downloaded from The Designer's Guide (www.designers-guide.com).
// Post any questions on www.designers-guide.com/Forum.
// Documentation on the model can be found at www.designers-guide.com/Modeling

`include "discipline.vams"

module varactor(p, n);
    parameter real c0 = 1p from (0:inf);        // nominal capacitance (F)
    parameter real c1 = 0.5p from [0:c0);       // maximum capacitance change from nominal (F)
    parameter real v0 = 0;                      // voltage for nominal capacitance (V)
    parameter real v1 = 1 from (0:inf);         // voltage change for maximum capacitance (V)
    inout p, n;
    electrical p, n;
    real q, v;

    analog begin
        v = V(p,n);
        q = c0*v + c1*v1*ln(cosh((v – v0)/v1));
        I(p, n) <+ ddt(q);
    end
endmodule
```

documentation of your model. You should also add a brief description to each parameter, and that description should include the units of the parameter if appropriate. The descriptions make it dramatically easier for someone to pick up and use your model, and for you to remember how to use the model when coming back to it after a period when you were not actively using it. The units serve to clarify the description.

After declaring the parameters you should declare the ports. Providing a description of each port as shown in Listing 17 on page 67 is generally a good idea, but not always necessary. In the model shown in Listing 30 the ports are undifferentiated and so do not merit a description. With the gate models, Verilog established a convention of placing the outputs first on the port list, with the primary inputs following the outputs, and the control inputs coming last. That convention is honored in this book. This is demonstrated in Listing 26 on page 85. In addition, when ports come in positive and negative pairs, they are given together with the positive port placed first, as shown in Listing 13 on page 61.

Finally, it is strongly recommended that you indent your code so that the extent of compound statements and scope are easily identified. Using 2-5 spaces for an indent level makes the indentation reasonably obvious without too much of a tendency to squeeze your code against the right margin. It is also helpful to add an extra blank line between logically separate blocks of code. In addition to the blank line, you should add a comment at the top of each block. and you should add comments to describe anything obscure in your model.

What's Next

This chapter introduced the Verilog-A language. The mixed-signal extensions for Verilog-AMS are discussed in the next. Neither is complete in its coverage. The intent is to introduce the language and present the concepts that it was built on rather than be a comprehensive user's guide for the language. More of the details of the language can be found in Chapter 5. Beyond that, the user's guide to your simulator or the language reference manual can serve as a comprehensive source of information about the language [28]. In addition, there are several books available that describe the digital part of the language in fuller detail [1,5,23,27].

As mentioned before, Verilog-A is the analog subset of Verilog-AMS. It is the part of the language that is suitable for implementation in a SPICE-class simulator. It is important to recognize that even those systems that can be adequately described using Verilog-A might benefit from Verilog-AMS. Verilog-AMS provides substantially more powerful and more efficient event driven modeling capabilities, and many purely analog systems are well modeled using an event-driven paradigm. So keep reading ...

4 Mixed-Signal Modeling

1 Mixed Signal Models

Both analog and digital functionality as well as the interaction between the two domains are described in mixed signal models. The Verilog-AMS language allows combining analog and digital behavior into a single model. That means that such a model can contain both mixed signal behavioral descriptions or it could instantiate a collection of analog, digital and mixed signal modules. In most cases these models have both analog and digital pins. But this is not always the case; occasionally a model looks digital in terms of its pin types, but analog behavioral descriptions are used inside. Or it may happen that a model looks purely analog when considering the type of its pins, but inside digital constructs are used, perhaps to speed up the model.

When using a top-down design methodology for developing a mixed-signal IC there are various needs for mixed-signal models. During the architectural phase, when blocks are represented abstractly, a single block often includes both complex analog and digital functionality. It is too early to divide this block into purely analog or digital sub-blocks in this phase. Verilog-AMS is used to describe this mixed-signal functionality. At this high level of abstraction the modeling is mostly done in such a way that the disciplines of the pins match. However, as the implementation process proceeds, the blocks are refined to the point where there may be multiple versions of the same block, and those different versions may have different disciplines for the same pin. This implies that as the different versions are used, conflicts may arise between the discipline of a net, and the pins connected to that net. In these cases, ***interface components*** (or connect modules as per the Verilog-AMS language definition) would be needed to resolve the conflict. As the design proceeds to lower levels of detail, these interface components are needed more often. Verilog-AMS provides mechanisms to allow these interface components to be inserted automatically based on the disciplines of the pins and the nets. In doing so, Verilog-AMS not only provides the ability to naturally describe mixed-signal behavioral models, but it also allows mixed-signal models to be built-up from purely analog and digital blocks, and those models to be freely interconnected with Verilog-AMS automatically performing the signal conversion.

The goal of this chapter is to introduce basic mixed signal behavioral and structural modeling techniques, and to help the user to understand the concepts of interface components and their automatic insertion. The following section gives an overview of modeling in the digital domain for the reader not familiar with Verilog-HDL. The understanding of basic digital constructs is a pre-requisite for the mixed signal modeling concepts presented in Section 3 and Section 4.

2 Modeling Discrete Behavior

2.1 Language Basics

The ability provided by Verilog-HDL for describing discrete behavior is fully contained as a subset of Verilog-AMS. That means that every Verilog-HDL model can be legally used in a Verilog-AMS context.

2.1.1 Disciplines

Consider the simple inverter of Listing 1.

LISTING 1 *Verilog-HDL description of an inverter.*

```
module inverter (q, a);
    output q;
    input a;
    wire a, q;      // digital net type (declaration optional)

    assign q=~a;
endmodule
```

The module header in Listing 1 looks similar to the examples provided for Verilog-A. In comparison to these Verilog-A module examples we notice that there is no discipline declaration provided for *a* and *q* (*wire* is not a discipline). Disciplines are a concept that is not part of Verilog-HDL. Disciplines first appeared as part of Verilog-A for continuous-time signals and Verilog-AMS extended the concept to also cover discrete-time signals, however it was made optional for these signals so that Verilog-HDL models can be used by a Verilog-AMS simulator without modification. The default discrete-time discipline is *logic*, which is defined in the *disciplines.vams* file and is shown in Listing 2.

LISTING 2 *The declaration of the discrete discipline 'logic'.*

```
discipline logic
    domain discrete;
enddiscipline
```

The user may define several discrete disciplines to distinguish between different logic families, different semiconductor processes, different supply voltages, etc. Nevertheless the inverter example shown in Listing 1 could be used directly without discipline definition and the discipline for a and q would default to *logic*. Listing 3 adds the explicit discipline declaration to Listing 1, which will later allow more control over the interface component insertion process.

LISTING 3 *The inverter of Listing 1 enhanced with a declaration of wire discipline.*

```
`include "disciplines.vams"

module inverter (q, a);
    output q;
    input a;
    wire a, q;      // digital net type (declaration optional)
    logic a, q;

    assign q = ~a;
endmodule
```

2.1.2 Wires

Now consider the details of the inverter module. The input and output statements are already known from Verilog-A.

```
    wire a, q;
```

defines that a and q are scalar wires. A ***wire*** is one type of a digital net. A scalar wire can carry one bit of information, and that bit can take one of the four values shown in Table 1 (5§2.5p164).

A wire is the logical representation within Verilog of a physical wire. As such, it can be connected to many things. In particular, there may be more than one thing driving it. An important aspect of the semantics of a wire is how it responds when driven by multiple outputs (drivers). If all of the drivers connected to a wire output a 0, the value of the wire is 0; if they all output a 1, then the value of the wire is 1. However, if the outputs produced by the drivers conflict, the wire resolves to x or unknown. The value x is interpreted as "either 0 or 1 or in a state of change". Any drivers that output a z

TABLE 1 *Verilog logic values.*

Name	Description	Literal Constant
0	Zero, low, or false.	0 or 1'b0
1	One, high, or true.	1 or 1'b1
x or *X*	Unknown or uninitialized.	1'bx
z or *Z*	High impedance (floating).	1'bz

(high impedance or disconnected) are ignored, unless all drivers output a *z*, in which case the wire resolves to a *z*.†

A wire is not the only type of net available in Verilog (5§2.5p164). There are also those that perform the equivalent of a *wired-or* or *wired-and* operation, those that are equivalent to connecting to the supply or to ground, and one, *trireg*, that mimics a net that holds its value due to charge storage.

It is also possible to declare a bus (a vector wire) by adding a range specification to a wire declaration (5§2.5p164). The range consists of the integer indices for the first and last members of the bus. For example,

```
wire [7:0] data;
```

declares an 8-bit bus, where the members can be accessed with *data*[*i*]; and *i* can range from 7 to 0.

2.1.3 Continuous Assignment

With the exception of the *trireg*, wires do not store values. Rather, they only transmit values that are driven on to them. To continuously transmit a value, it must be continuously driven. One way to do this is with a ***continuous assignment statement*** (5§7.4.2p213), which is the last new statement given in Listing 1.

```
assign q = ~a;
```

indicates that *q* is driven at all times to the value that is the inverse of *a*: the state of *q* changes directly with any change of *a*. The operator '~' is the bitwise invert operator, see (5§4.1p172) for a list of all operators available in Verilog-A/MS.

† In certain cases, resolution rules beyond the scope of this book are triggered that cause the result to be somewhat different. For details, see the Verilog and Verilog-AMS LRMs [16,28].

2.1.4 Processes and Registers

Continuous assignments by themselves only consist of a single statement, and so can only represent relatively simple behavior. For example, it is easy to implement a logical combination of values, but more difficult to describe operations that depend in complicated ways on the past value of signals. For these, Verilog adds the concept of processes and registers.

Processes are independently operating threads of control, and so, in a sense, a continuous assignment statement represents a process. However, it is the *initial* and *always* processes that are of primary interest at the moment. A module may have any number of *initial* or *always* processes; all of which start up as the simulation begins. Each consists of a single statement. The ***initial*** processes will execute their statements once and then terminate. The ***always*** processes execute their statements repeatedly; they never terminate (5§7.1p209). Both are used to implement the clock generator of Listing 4.

LISTING 4 *Simple clock generator.*

```
`timescale 1ns / 1ps

module clock_gen (clk);
    parameter cycle = 20;       // clock period (ns)
    output clk;
    reg clk;

    initial clk = 0;

    always #(cycle/2) clk = ~clk;
endmodule
```

This module has a single *initial* process that initializes the value of *clk*. *clk* is a new type of variable called a ***reg***, which is short for ***register***. A register is nothing more than a variable that can hold a logic value (0, 1, *x*, or *z*). Like all variables in Verilog, it retains its value until explicitly changed, but since registers hold bit values, they act like a hardware register (and hence the name).

The *always* process repeatedly executes the statement

```
#(cycle/2) clk = ~clk;
```

This statement has two parts to it. The first is a delay specification. The construct #*x* tells Verilog to delay for *x* units of time before proceeding (5§7.6.1p216). A unit of time is defined by the first statement in Listing 4,

```
`timescale 1ns / 1ps
```

This statement gives two values, the first is the duration of a unit of time and the second represents the resolution of time. This statement indicates that one unit of time is 1 ns, and fractional units of time are rounded to multiples of 1 ps (5§1.4p151).

The second part of the statement is a simple assignment that sets the new value of *clk* to the inverse of its current value. Thus, in one pass through the *always* process in this example, execution pauses for *cycle*/2 ns and then inverts the value of *clk*, and this action is repeated forever with the result being that the value of *clk* alternates between 0 and 1 with a period of *cycle* nanoseconds. If the default value of 20 is used for *cycle*, the resulting period is 20 ns and the frequency is 50 MHz.

A continuous assignment could now be used to take the value of *clk* and use it to drive the output, but this is not necessary. A register can act as an output. And so, in this example the output of the clock generator is declared to be the value of the *clk* register. Of the variables, only registers have the ability to act as outputs; integers and reals do not.

2.1.5 Timing Control

The delay construct used in Listing 4 is one of three types of timing control constructs provided by Verilog (5§7.6p216), which are listed in Table 2. Listing 5 demonstrates another type: ***edge triggering***.

TABLE 2 *Verilog timing control constructs.*

#<*delay*>	Simple delay
@(<*signal*>)	Edge-triggered timing control
wait(<*expr*>)	Level-sensitive timing control

LISTING 5 *Edge triggered d-flip-flop.*

```
module dff (q, d, clk);
    output q;
    input d, clk;
    reg q;

    always @(clk)
        if (clk)
            q = d;
endmodule
```

The construct @(*clk*) tells Verilog to wait for an event to occur on *clk* before proceeding (5§7.6.2p216). An ***event*** is defined as a change in value. Once an event occurs, the *if* statement is executed. In this model it determines if a rising edge has occurred, and if so the output *q* is set to the current value of the input *d*. Triggering on only the rising or falling edge is common, in which case the *posedge* or the *negedge* keywords are used to specify which type of events are of interest. Thus, the always block could be simplified to

```
always @(posedge clk)
    q = d;
```

The third type of timing control is the *wait* statement, illustrated in the example of Listing 6.

LISTING 6 *A two input latch.*

```
module latch2 (out, d1, d2, en);
    output [1:0] out;
    input d1, d2, en;
    reg [1:0] out;

    always @(d1 or d2)
        wait (!en) begin
            out[0] = d1;
            out[1] = d2;
        end
endmodule
```

The *wait* is ***level triggered***; it pauses execution of the process until its argument is true (5§7.6.3p218). In this case it waits for !*en* to go to 1. The '!' operator returns the logical negation of its argument *en*, and so in this case the process waits for *en* to become 0 (5§4.1p172). Level triggering differs from event triggering if the desired condition is present when the delay construct is executed. In that case, level triggering provides no delay, whereas event triggering will wait for the next time that the condition becomes true.

2.1.6 Vectors

Listing 6 also demonstrates the use of busses or bit vectors. With

```
reg [1:0] out;
```

register *out* is declared as a vector with two members, the first having an index of 1 and the second having an index of 0 (5§2.2p155). This declaration is largely duplicated in the output specifier,

```
output [1:0] out;
```

but it need not be, it is sufficient to simply declare *out* as being an output with

```
output out;
```

The members of the bit vector are assigned values with the statements

```
out[0] = d1;
out[1] = d2;
```

The individual members were accessed by placing the index of the member in brackets, a process called ***bit selection***. They could also have been assigned in mass using

```
out = {d1, d2};
```

Here the braces create a vector by concatenating the list of arguments within the braces, and then that vector is assigned to *out* (5§2.1.4p154).

2.1.7 Procedural Blocks

The latch model of Listing 6 introduces one more very important concept: procedural blocks. A ***procedural block*** is a compound statement that starts with the *begin* keyword and finishes with the *end* keyword (5§7.2p209). From the outside it appears as a single statement, but once executed each of the statements it contains is executed in order, with a statement executing after the statement that precedes it completes. When the last of its statements completes, the procedural block itself completes. In this example, the procedural block consists of

```
begin
    out[0] = d1;
    out[1] = d2;
end
```

In this block, *out*[0] is assigned first, and *out*[1] is assigned after *out*[0]. While *out*[1] is assigned after *out*[0], it is important to recognize that from the perspective of the system being simulated, no time passes between when the two statements execute. System time advances only when evaluating the various timing control constructs (#, @, *wait*).

2.1.8 Concurrent Blocks

Procedural blocks represent a compound statement where the component statements execute sequentially, or in series. Verilog also provides a compound statement where the component statements execute concurrently, or in parallel: a *fork-join* block (5§7.3p211).

Example:

```
fork
    #5      a = 1;
    #10     a = 0;
    #15     b = 1;
join
```

In this example, all three statements are launched at the time the block is executed, and the block completes after all of its component statements have completed. The value of *a* is set to 1 after 5 time units and then returned to 0 after another 5, and the value of *b* is set to 1 and the compound statement completes after another 5 time units, or 15 units after the block started executing as shown in the time line below.

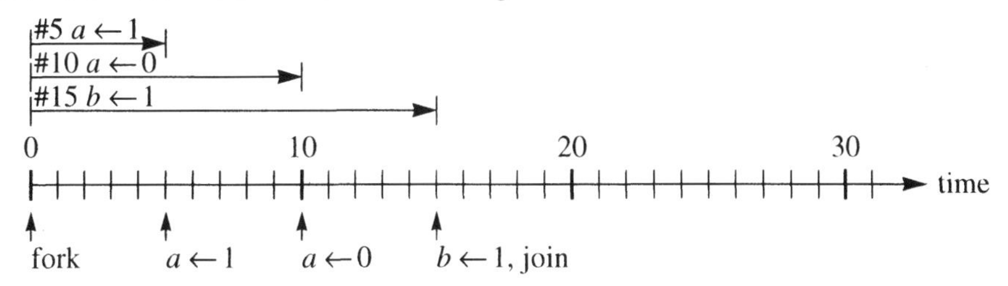

Contrast this with the same set of statements contained in a procedural block.

```
begin
    #5      a = 1;
    #10     a = 0;
    #15     b = 1;
end
```

In this case, the delays accumulate as shown in the time line below.

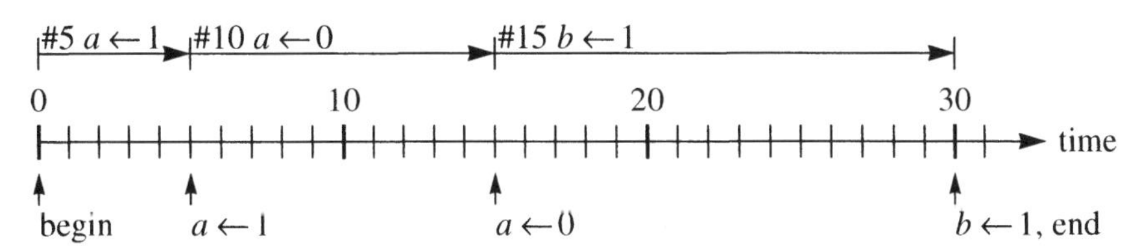

The *fork-join* structure is especially useful for producing test bench stimuli.

2.1.9 Delayed Assignment

In the two forms of assignment presented so far the assignment occurs at the same time as the evaluation of the expression on the right side of the assignment. In hardware, there is typically a delay between these two. Adding delay before the assignment can be done, but it delays both the evaluation and the assignment. Verilog allows delay to be inserted directly into the assignment statement itself. This is referred to as ***inter-assignment delay*** (5§7.6.4p218). There are two ways in which this can occur. The first inserts a simple delay,

Example:

```
a = #5 b;
```

In this example, the right-hand side is evaluated, the delay occurs, and then *a* is updated and the statement completes. The same can be done with event delays.

Examples:

```
q = @(posedge clk) d;
```

Swapping of two values can be preformed without a temporary variable using a concurrent block and delayed assignment.

```
fork
    a = #1 b;
    b = #1 a;
join
```

Without the delay, a race condition would occur, making the final outcome indeterminate.

2.1.10 Non-Blocking Assignment

The assignment found in initial and always processes so far have been blocking assignments. They are called that because adding the delay acts to pause or 'block' the execution of the process. An alternative type of assignment is the non-blocking assignment (5§7.4.1p212), demonstrated below.

Example:

```
always @(negedge clk) begin
    a <= ~b;
    b <= ~a;
end
```

If this code employed blocking assignments, once a falling edge occurred on *clk* the first assignment statement would update *a* before the second assignment evaluated its right side, with the end result that *b* returns to its original value. However, when more than one non-blocking assignment is executed at the same time, the right sides of all of them are evaluated before any left side is updated. As a result, they all appear to occur concurrently. Thus, if the non-blocking assignments occur at time t, then the value of the arguments at time t^- is used to produce the updates, which occur at time t^+. For the example above, a falling edge on *clk* triggers the execution of the procedural block causing the right side of each non-blocking statement to be evaluated to form $\sim a^-$ and $\sim b^-$, then the assignment targets are updated to $b^+ = \sim a^-$ and $a^+ = \sim b^-$, and finally the process blocks and waits for the next falling edge on *clk*. The end result is that *a* and *b* are both swapped and inverted on every falling edge of *clk*.

Delay can be added to non-blocking assignments in the same way it was with blocking assignments. However, the delay does not act to block the execution of the process in which the assignment resides. Instead, the right side expression is evaluated, the result is set aside temporarily, and the update of the target is scheduled into the future and the statement completes. Later, once the specified delay passes, the target is updated to the saved result.

2.1.11 Delayed Continuous Assignment

Delay can also be added to continuous assignments, but in this case the delay is specified before the assignment target.

Example:

```
assign #5 a = b;
```

2.2 Integers and Reals

Integer and real values provide the user with the ability to write models at a higher level of abstraction than if only simple wire and register types were used, which are more closely tied to the actual hardware implementation. Verilog-HDL provides integers and reals as "abstract data types", as does Verilog-AMS. In addition to supporting integers and reals as constants and variables, Verilog-AMS also allows integer and real valued discrete-event signals to be transported across the module boundaries with the *wreal* wire type. It is important to stress here that the wreals, as well as the integer- and real-valued variables described here, are all associated with the discrete-event kernel. Comparing the value of a wreal wire with the voltage potential of an *electrical* node reveals that the numerical value of both is real, meaning that the signals can take on any real value, but the wreal wire can only change its value at discrete points in time and is otherwise constant while the potential of an *electrical* node can change continuously with time. Thus, the wreal wire is discrete in time whereas the potential of an *electrical* node is continuous in time.

As an example of the use of integers in a module, consider the counter shown in Listing 7. At every rising edge of *clk* the integer variable *count* is incremented by one until it equals the value of *maxcount*, which has the default value of 9. At this point the output register *out* becomes high for the length of the period of a clock cycle. In this way a transition on the output occurs for every *maxcount* transitions on the input.

LISTING 7 *Behavioral model of a counter.*

```
module counter (out, clk);
    parameter maxcount = 9;            // the number of input pulses per output pulse
    input clk;
    output out;
    reg out;
    integer count;

    initial begin
        out = 0;
        count = 0;
    end

    always @(posedge clk) begin
        count = count + 1;
        if (count == maxcount) begin
            out = 1;
            count = 0;
        end else if (count == 1)
            out = 0;
    end
endmodule
```

Listing 8 contains a frequency measurement module. It calculates the frequency of a digital input clock based on the time difference between two rising edges. The variables *last_time*, *current_time* and *freq* are declared as reals. The frequency value is retained within the module, but it can still act as a useful output to the user as it can be read directly by the debug or waveform tools. The *$realtime*† system function returns the current digital simulation time. It provides the real representation of the simulation time in the time units defined by the *`timescale* directive. This time is converted to seconds by multiplying the return value of *$realtime* with the time unit setting for this module.

† In early versions of Verilog-A, *$realtime* was used to return the current simulation time as a real value in seconds. When Verilog-HDL and Verilog-A were combined into the Verilog-AMS standard, the Verilog-HDL behavior of this system function was kept and the Verilog-A version of this functionality was renamed to *$abstime*.

LISTING 8 *Digital frequency measurement.*

```
`timescale 1ns / 1ps

module freq_meas (clk);
    input clk;
    real last_time, current_time, freq;

    initial begin
        last_time = 0.0;
        freq = 0.0;
    end

    always @(posedge clk) begin
        current_time = $realtime;
        if (last_time > 0.0)
            freq = 1.0e9 / (current_time – last_time);
        last_time = current_time;
    end
endmodule
```

2.2.1 Real Wires

The real values used in Listing 8 are variables. The usage of variables is limited to within the module context only; and unlike registers, real and integer variables cannot be used as ports. The transfer of real values between modules is cumbersome in Verilog-HDL. The real wire type *wreal* was added to Verilog-AMS to make it easy to pass both integer and real values between modules. The example in Listing 9 shows how a *wreal* can be used to describe the behavior of a DAC. This module is very similar to the Verilog-A DAC found in Section 11 of Chapter 3 except for the *wreal* output. Since a real variable cannot act as a port, it is necessary to use a continuous assignment to drive the wreal port.

3 Modeling Mixed-Signal Behavior

Verilog-AMS not only combines the syntax elements of Verilog-A and Verilog-HDL, but it allows both within a single module. For the analog modeler it adds powerful elements for efficiently describing event-driven behavior. It can both shorten the model development time and reduce the simulation time if used in the right way. A good example of the simulation speed-up that can be achieved by leveraging mixed-signal techniques is a phase-locked loop (PLL) based frequency synthesizer like the one shown in Figure 1.

LISTING 9 *Verilog-AMS model for a digital-to-analog converter with wreal output.*

```
`timescale 1ns / 1ps
`include "disciplines.vams"

module dac (out, in, clk);
    parameter integer bits = 8 from [1:24]; // resolution (bits)
    parameter real fullscale = 1.0;           // output range is from 0 to fullscale (V)
    parameter real td = 0.0;                  // delay from clock edge to output (s)
    parameter integer dir = 1 from [-1:1] exclude 0;
                                              // +1 triggers on rising clock edge, -1 on falling
    output out;
    wreal out;
    input [0:bits-1] in;
    input clk;
    logic in, clk;

    real result, aout;
    integer weight;
    integer i;
    parameter integer idir = (dir == 1 ? 1 : 0);

    always @(clk) begin
        if (clk == idir) begin
            aout=0.0;
            weight = 2;
            for (i=bits-1; i>=0; i=i-1) begin
                if (in[i]) aout = aout + fullscale / weight;
                weight = weight * 2;
            end
            result = #(td / 1.0E-9) aout;
        end
    end

    assign out = result;
endmodule
```

A PLL-based frequency synthesizer generates a high frequency signal that is an exact multiple of a lower frequency reference signal. The voltage controlled oscillator (VCO) produces the high frequency output signal (*out*). Its output frequency is made an exact multiple of the reference frequency (*ref*) by using feedback. The reference frequency is compared to a divided down version of the VCO output frequency (*fb*) produced by the frequency divider (FD). Any difference in the phase or frequency between the two is sensed by the phase-frequency detector (PFD) and converted to an

FIGURE 1 *Block diagram of a PLL frequency synthesizer.*

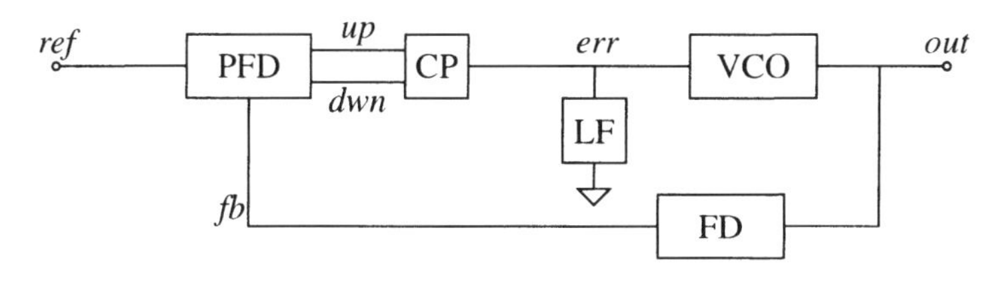

error signal (*err*) by the charge pump (CP) and loop filter (LF) that is used to adjust the VCO control signal so as to eliminate the frequency error.

The VCO, charge pump and loop filter are all circuits that exhibit analog behavior whereas the frequency divider and phase-frequency detector are purely digital. If we were not allowed to construct mixed-signal models, the first three blocks would be represented with Verilog-A and the latter two with Verilog-HDL. In order to connect the two types of models, some sort of interface components that converts signals between the continuous and discrete time domains would be needed at nodes *up*, *dwn*, and *out*. Such an approach would be relatively inefficient, as both the charge pump and the VCO, while analog in nature, can be more efficiently modeled using the analog event-driven capabilities of Verilog-AMS. Consider the VCO. It produces the highest frequency signal in the design and therefore, of all the components, most strongly affects the simulation time. Assume the Verilog-A model given in Listing 22 on page 74 is used for the VCO. It produces a sinusoidal output signal and the model forces the simulator to take at least 10 time steps per period to achieve sufficient accuracy, which will cause the simulation of the VCO to be expensive. However, even if the output were converted to a square wave, at least 4 time points would be needed per output period, with a more likely number being 10-20. Either way, the VCO is expensive to simulate in the continuous-time domain. This is where Verilog-AMS can make a big difference. Modeling the output of the VCO in the discrete-event domain and the input circuitry of the VCO in the analog domain would result in many fewer continuous-time points being needed. Just such a model for the VCO is shown later in Listing 13 on page 120. Until then, the mechanics of constructing mixed-signal models are presented. This will allow us to both eliminate the need for external interface components and allow us to build models such as the VCO that naturally span the two domains.

3.1 Analog and Digital Contexts

The module header and declaration portions of the module are used for both analog and digital elements. The behavioral description portions are separate. The analog behavior is described in the analog process recognizable by the *analog* keyword that is known from Verilog-A. Digital behavior is described outside of this process.

It should be understood that by analog we mean 'continuous time' and by digital we mean 'discrete event'. In reality, analog is not synonymous with continuous time. Analog really means continuously valued, and it is possible to have continuous-value discrete-event models. However, with regards to mixed-signal simulation, common usage associates the term 'analog' with the continuous-time kernel and semantics. Similarly, digital is used to refer to the discrete-event kernel and semantics even though it is possible to use them to represent continuous-value, and hence analog, discrete-event models.

From within an analog process it is permissible to read but not modify the data associated with discrete processes. In particular, from an analog process one can access the values of discrete-event wires, events, and registers and other variables. This means, analog behavior can be modeled that is dependent on the status and events of the discrete domain portion of the module. Similarly, the digital behavior described outside of the analog process can be dependent on signals, values, and events associated with analog processes.

Although analog and digital variables are declared identically, they are owned by either the analog or the digital context. One cannot distinguish from the declaration

```
real x;
```

whether it is owned by the analog or by the digital context. However, Verilog-AMS only allows write access to one domain. If it is the target of an assignment in an analog process, it is said to be owned by the continuous kernel. Conversely, if it is assigned a value in a discrete process (initial or always) it is owned by the discrete kernel. In other words, the context in which the assignment to a variable takes place owns the variable. Thus an assignment in an *always* block

```
always @(posedge clk)
    begin
        ...
        a = 1;
    end
    ...
```

would mean that the real variable *a* is owned by the digital kernel. Other assignments to *a* would be allowed in the digital context like in other *always* or in *initial* blocks, but assignments to *a* in the analog process would be illegal. However read-access to *a* would be allowed in both contexts. On the other hand if the assignment or write-access to the variable is made in the analog context like the real variable *reff* in

Listing 11 shown later, the variable is owned by the analog kernel. Even though it is possible to read it in the digital context it could not be changed from there.

3.2 From Digital to Analog

If the analog behavior is dependent on digital conditions, the analog process must explicitly access information from the digital context. This information can be accessed in two different ways: the state of wires and variables from the digital context can be read when their value is needed, or a certain change of the digital state can be used to trigger changes in the analog behavior.

3.2.1 Digital Access on Demand

The model for a digitally controlled ideal switch is given in Listing 10.

LISTING 10 *Digitally controlled switch.*

```
`include "disciplines.vams"

module switch (p, n, s);
    input s;
    output p, n;
    logic s;
    electrical p, n;

    analog begin
        if (s)
            V(p, n) <+ 0.0;
        else
            I(p, n) <+ 0.0;
    end
endmodule
```

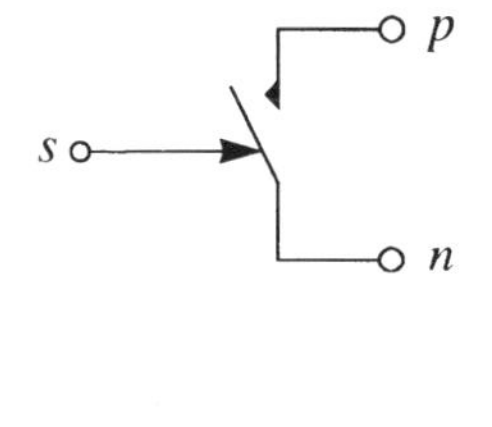

In the module header and declaration portion both analog and digital inputs and outputs are declared. The analog process is executed at each analog simulation time step. The switch is on or off dependent on the status of *s*. If the value of *s* is 1 the switch is closed, otherwise the switch is open. This mixed-signal model does not contain any digital behavior; it just reads a digital state needed to maintain the analog behavior. Thus there are no digital behavioral statements in the module.

3.2.2 Analog Sensitivity to Digital Events

Instead of just reading the discrete domain data it is often more efficient to make the analog behavior sensitive to events in the discrete domain. That way the analog can react at the instant when a digital value changes. Listing 11 shows an example where this modeling style is used to describe a digitally controlled non-ideal switch.

LISTING 11 *Digital controlled switch with internal resistance.*

```
`timescale 1ns / 1ps
`include "disciplines.vams"

module switch (p, n, s);
    parameter real ron = 10.0 from (0:inf);          // on resistance (ohms)
    parameter real roff = 100.0M from (ron:inf); // off resistance (ohms)
    parameter real td = 0.0;                          // delay time (s)
    parameter real tr = 20n;                          // rise time (on →off) (s)
    parameter real tf = 20n;                          // fall time (off →on) (s)
    input s;
    logic s;
    electrical p, n;

    real reff;

    analog begin
        @(posedge s) reff = ron;
        @(negedge s) reff = roff;
        @(initial_step) reff = (s ? ron : roff);

        I(p, n) <+ V(p, n) / transition(reff, td, tr, tf);
    end
endmodule
```

During the initial analog simulation step the target resistance value *reff* of the switch is set dependent on the digital input state *in*. The state of *in* is read as in the previous example. Afterwards, changes in the value of *reff* result from events on *in*. When digital events occur that are being monitored by the continuous-time kernel, mixed signal synchronization results. The Verilog-AMS language reference manual [29] clearly defines this synchronization algorithm, thereby ensuring the same behavior of the model when used with different simulators built according to the Verilog-AMS standard. When these digital-to-analog events occur the synchronization algorithm guarantees that the digital change is noticed by the continuous-time kernel immediately.

Figure 2 illustrates how the synchronization algorithm works for the case of the

```
@(posedge s)
```

statement in the analog block:

1. The continuous-time kernel has solved time point *a* and the discrete-event driven kernel has processed all the digital events up to the digital time point *b*.

FIGURE 2 *Digital-to-analog synchronization*

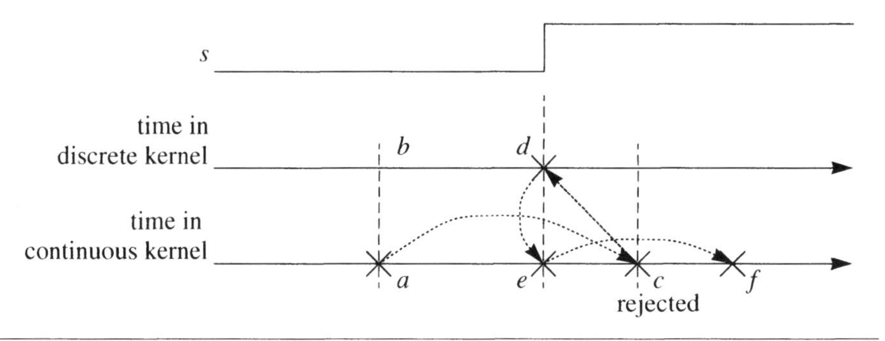

2. The analog kernel chooses the analog time point *c* as its next time point to solve according to its built-in time step mechanism and solves the analog system at this point.
3. The digital kernel is now enabled to proceed forward to evaluate the digital events up to a digital time point equivalent to *c*, but at the time *d* the value of *s* changes state. Because there is an analog sensitivity to this event, it causes the discrete kernel to stop at *d* and to hand over the simulation control to the continuous-time kernel.
4. The already calculated analog time point *c* is rejected, because it did not take the analog reaction to the digital event at *d* into account.
5. The continuous kernel solves the analog system at *e*. The time of *e* is chosen to exactly equal the time of *d*. The analog solution at this point accounts for the change in *s*. In our example this means that *reff* is set to *ron*. Thus the digital event is recognized immediately and leads to changed analog behavior.
6. The analog kernel again takes the lead by advancing time to point *f*, where it solves the analog portion of the circuit. It then passes control back to the digital kernel in order to allow it to catch up.

The transition function is used to smooth the effect of the abrupt change in the value of *reff*. This is a recommended practice as the continuous-time kernel often times does not react well to abrupt changes. In this case, the transition function ensures that the analog behavior changes continuously. But more than that, it reduces the variation in the behavior of the system that would occur as a result of variations in the size of the time step that either immediately precedes or follows the abrupt change. The transition function defines exactly the beginning, the end, and the rate of change within the transition (5§4.6.4p180).

3.3 From Analog to Digital

The values of the analog domain can be accessed from within the discrete context just as in the opposite case just described. And again we differentiate between the two major ways of accessing the data: access by demand, meaning the reading of the value of analog signals or variables when the data is required on the digital side; and the detection of analog events like threshold crossings and the transformation of these events into the digital domain.

3.3.1 Analog Access on Demand

Listing 12 shows a discrete-event analog to digital converter. It is similar to the one given in Listing 26 on page 85, but rewritten to employ the faster discrete-event kernel. The conversion takes place on each rising edge of the clock. At that point the analog input signal is read. Verilog-AMS does not require the analog kernel to place a point at the instant when the analog value is sensed, and so the value is likely interpolated.

This represents an important difference between digital access on demand and analog access on demand. If an analog process is sensitive to a digital signal, then it is evaluated at the point when the digital signal changes. However, the reverse cannot be true for discrete processes when they are sensitive to continuous-time signals as these signals can change continuously. If it is desired to force the analog kernel to place a time point at the instant when the converter samples the input signal in order to increase the accuracy of the samples, add the following lines to the model.

```
analog @(posedge clk)
    ;
```

In Listing 13 is a Verilog-AMS description of a VCO in which the analog input voltage controls the frequency of an output clock. This mixed signal VCO can be used to significantly improve the simulation run-time of the PLL mentioned earlier. The analog input voltage is sampled at half the period of the output clock. The control voltage for the output frequency is the voltage difference between the inputs *ps* and *ns* (short for positive sense and negative sense). The parameter *f0* is the center frequency of the VCO output. It is the frequency measured at the output when the input differential voltage is zero. The *kvco* parameter determines the voltage-to-frequency gain. The output clock generation is a modification of the simple clock generator shown in Listing 4. The clock frequency is not a constant anymore. Now it is dependent on the input voltage. The clock output *out* starts at logic low at time zero and is inverted after every half period of the desired output frequency. With

```
vin = V(ps, ns);
```

LISTING 12 *Verilog-AMS model for an N-bit analog-to-digital converter.*

```
`include "disciplines.vams"
`timescale 1ns / 1ps

module adc (out, in, clk);
    parameter integer bits = 8 from [1:24];  // resolution (bits)
    parameter real fullscale = 1.0;           // input range is from 0 to fullscale (V)
    parameter real td = 0;                    // delay from clock to output (ns)
    input in, clk;
    output out;
    voltage in;
    reg [0:bits-1]out;
    reg over;
    real sample, midpoint;
    integer i;

    always @(posedge clk) begin
        sample = V(in);
        midpoint = fullscale/2.0;
        for (i = bits - 1; i >= 0; i = i - 1) begin
            over = (sample > midpoint);
            if (over)
                sample = sample - midpoint;
            sample = 2.0*sample;
            out[i] <= #(td) over;
        end
    end
endmodule
```

the differential input voltage is sampled every half period and the value stored in the digital real variable *vin*. Afterwards *vin* is used to calculate the length of half the period of the desired output frequency.

3.3.2 Digital Sensitivity to Analog Events

Whereas the analog values are simply read when a new analog value is needed in the digital context, digital behavior can also be triggered when a certain condition occurs on the analog side. Such analog events could be the crossing of a voltage threshold or an analog timer event for instance. To make a digital block sensitive to an analog event, the analog event expression is put directly into the digital block instead of a digital event expression. The following *always* process uses the *cross* function to detect when the voltage at node *in* crosses 2.5 V in the positive direction.

LISTING 13 *Mixed-signal voltage controlled oscillator.*

```
`timescale 1ns / 1ps
`include "disciplines.vams"

module vco (out, ps, ns);
    parameter real f0 = 100k from (0:inf);  // center frequency (Hz)
    parameter real kvco = 10k;              // gain (Hz/V)
    input ps, ns;
    output out;
    electrical ps, ns;
    reg out;
    logic out;
    real vin;

    initial out = 0;

    always begin
        vin = V(ps, ns);
        #(0.5e9 / (f0 + kvco * vin))
        out = ~out;
    end
endmodule
```

```
    always @(cross(V(in) - 2.5, 1)) begin
        dreg = ~dreg;
    end
```

When this occurs, the digital register *dreg* is inverted. This example could also use the *above* event function to detect the threshold crossing.

```
    always @(above(V(in) - 2.5)) begin
        dreg = ~dreg;
    end
```

The behavior would be similar to the example where the *cross* function is used, with the exception that the *above* function also produces an event if *V(in)* is above the threshold during the initialization phase of the simulation (during a DC or IC analysis) whereas the *cross* function only produces events at the actual threshold crossings.

The crossing point (*a* in Figure 3) is found by the analog solver. It then places an evaluation point, *b*, within the interval defined by the tolerances associated with the *cross* or *above* function. The analog evaluation point will not coincide precisely with the threshold crossing because the representation of time used by the analog kernel has only finite precision. Instead, the evaluation point is placed just beyond the threshold crossing. In this way, any conditionals within the analog behavioral description will detect the crossing. The discrete kernel also represents time using finite precision, with the resolution being given in the `*timescale* directive. Therefore, the time of the digital evaluation point, *c*, will be displaced somewhat from both the time of the crossing, and from the time at which the analog kernel places its evaluation point. In this case, the digital event is scheduled at the closest digital evaluation point (point *c* in Figure 3) exactly at or earlier than the time point where the analog event occurred.

FIGURE 3 *Analog-to-digital synchronization.*

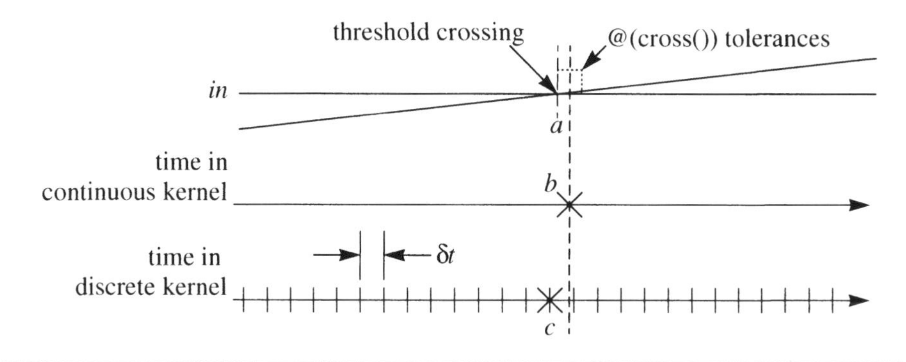

A typical behavioral element that uses analog-to-digital events is the comparator module shown in Listing 14. The input of this module is an analog differential voltage. The output is a logic signal. The output changes when the input voltage difference crosses a certain voltage threshold. The parameter *hyst* provides the capability to add a hysteresis between the lower and the upper thresholds. The example uses the *above* statement to ensure a correct setting of *out* from the beginning of simulation.

4 Structural Verilog-AMS

Verilog-AMS is more than just a mixed-signal behavioral modeling language. In Chapter 3, focusing on the analog subset Verilog-A, the capabilities of the language for structural descriptions were explained (3§2p41). The assembly of Verilog-A instances in a structural description or netlist and the mechanism for overwriting of the instance parameters is already known. This structural use of Verilog-A and its capability to mix these structural descriptions with behavioral descriptions within one and the same module is fully compatible with Verilog-AMS. Verilog-AMS allows the instantiation of analog Verilog-A, digital Verilog-HDL and mixed-signal Verilog-AMS modules. It also allows connection of these instances with each other and the mixture of netlist structure and mixed-signal behavior within a single module.

LISTING 14 *Comparator with logic output*

```
`timescale 1ns / 1ps
`include "disciplines.vams"

module comparator (out, p, n);
    parameter real offset = 0;                 // Offset voltage (V)
    parameter real hyst = 0.0 from [0:inf);    // Hysteresis (V)
    inout p, n;
    output out;
    electrical p, n;
    logic out;
    reg out;

    parameter real thrlo = offset – 0.5*hyst; // Lower threshold voltage (V)
    parameter real thrhi = offset + 0.5*hyst; // Upper threshold voltage (V)

    always @(above(V(p, n) – thrhi))
        out = 1;

    always @(above(thrlo – V(p, n)))
        out = 0;
endmodule
```

It is important to note that both the digital Verilog-HDL as well as the analog Verilog-A language support their own subset of the Verilog-AMS structural syntax. Especially in the case of Verilog-HDL it could be a good idea to limit the use of Verilog-AMS to this subset if all instances in the netlist are purely digital. This has the advantage that if need be, the module could be read by a Verilog-HDL simulator without modification. This is useful if a block is reused in a purely digital design, or in cases where purely Verilog-HDL compatible descriptions of the design are needed to share with members of the design team that are not analog savvy.

4.1 Connecting Analog and Digital

Verilog-AMS allows one to directly connect analog and digital ports of instances in a netlist or structural description. As a simple example, Listing 15 connects the digital clock generator introduced in Listing 4 directly to a Verilog-A resistor. Note that the net *x* is not explicitly declared in the netlist. It is the task of Verilog-AMS's discipline resolution algorithm to determine which domains and disciplines should be used for the undeclared nets.

Once the domains and disciplines are resolved in the design, it may happen that ports from different domains are connected to the same node. It is at these nodes that connect modules are automatically inserted in order to act as translators between the two

LISTING 15 *Netlist that connects analog and digital modules.*

```
`include "disciplines.vams"

// the resistor
module res (p, n);
    parameter real r = 100k from (0:inf);    // resistance (Ohms)
    inout p, n;
    electrical p, n;

    analog I(p, n) <+ V(p, n) / r;
endmodule

// the netlist
module top;
    clock_gen #(.cycle(1)) clock_gen1 (x);
    res #(.r(50k)) r1 (x, gnd);
endmodule
```

domains. Connect modules are also referred to as ***interface elements*** or ***interface components***. Users have full control over which connect modules are used and can write their own connect modules using the full power of the Verilog-AMS language. Connect modules even provide access to some capabilities that are not available in normal modules. Discipline resolution, connect module insertion, and the connect modules themselves are part of the Verilog-AMS language and will be discussed in the remainder of this chapter.

4.2 Discipline Resolution

To understand the automatic connect module insertion mechanism we must first understand the discipline resolution algorithm. The simulator resolves the net disciplines in several steps.

Initially the disciplines for the nets contained entirely within the scope of a module are determined. If undeclared, a net will take the discipline of whatever is connected to it. Within a module a net may be connected either to components instantiated from within the module or to behavioral descriptions given in the module. Behavioral access to continuous-time nets is only allowed for nets that have declared disciplines; however it is common for discrete behavioral models to access the values of wires and registers directly, without an implied discipline. In this case, the *`default_discipline* directive is used to determine which discipline should be used for the net. The *`default_discipline* directive allows pre-existing digital Verilog-HDL modules to be used without needing to add discipline declarations for all signals. Consider the following example.

```
`include "disciplines.vams"
`default_discipline logic

module test (out, in);
    output out;
    input in;
    reg out;

    always @(in) out = in;
endmodule
```

The *`default_discipline* directive causes both the register *out* and the wire *in* to be implicitly declared with a discipline of *logic*, just as if the following line had been found in the module.

```
logic in, out;
```

Of course, the default discipline could also be set to a user defined discipline, unlike in this example where the pre-declared discipline *logic* is used as the default. Simulators may also provide an option to set the default discipline for the whole design at once at the command line or in an options file.

In the second step of the discipline resolution algorithm, the out-of-module discipline declarations are taken into account. Such out-of-module discipline declarations are a method for coercing the mixed-signal discipline resolution from outside a module. For example, a user may want to force a particular net having no discipline to be declared *logic* to prevent it from becoming analog during the later steps of the discipline resolution algorithm. The discipline declaration statement

```
logic top.a73.x
```

in the netlist coerces the discipline of the domainless net *x* located within the instance *a73* that is instantiated in the top-level module *top* to the discipline *logic*.

After considering the out-of-module discipline coercions, mixed-signal discipline resolution resolves all nets that still do not have assigned disciplines. For this step Verilog-AMS provides two methods: the basic or default method and the detailed resolution method. However, it does not provide a way of selecting which method is used. This must be done using direct manipulation of the simulator options.

4.2.1 Basic Mixed Signal Discipline Resolution

Recall that a node is an electrically infinitesimal point of interconnection and a net is the name used for a node within a particular module. Thus, a node can consist of several nets, each existing in different modules and perhaps each with their own discipline declarations. These disciplines must be resolved to a single discipline on those nets where connect modules (interface components) are needed in order to determine which connect module to use.

With the default method of discipline resolution the nets of both discrete domain disciplines as well as continuous (analog) domain disciplines inherit their disciplines upward through the design hierarchy. That means a net with no discipline declared in an upper level of the design hierarchy (closer to the root of the hierarchy or the top-level module) will inherit its discipline from the ports (or nets) that connect to it from lower in the design hierarchy, and so on (5§2.4.4p162). In the case where there are multiple ports from a lower level connected to an undisciplined net the following resolution rules apply:

1. The net takes the discipline of the lower level ports if all the ports connected to the net share the same discipline.
2. The user must provide a resolution rule in the case where several compatible disciplines of the same domain (either discrete or continuous) meet each other. For disciplines to be compatible they must share the same domain (5§2.4.1p161). In addition, for continuous-time disciplines the natures must derive from the same base natures. All discrete disciplines are naturally compatible with each other. This case is illustrated in Figure 4. Here a net x without a declared discipline connects to a port with the discipline *logic_a* with a port having the discipline *logic_b*, where *logic_a* and *logic_b* are assumed to be discrete disciplines and so are compatible. In this case the user must define how the discipline conflict is resolved.

FIGURE 4 *Resolution for compatible disciplines.*

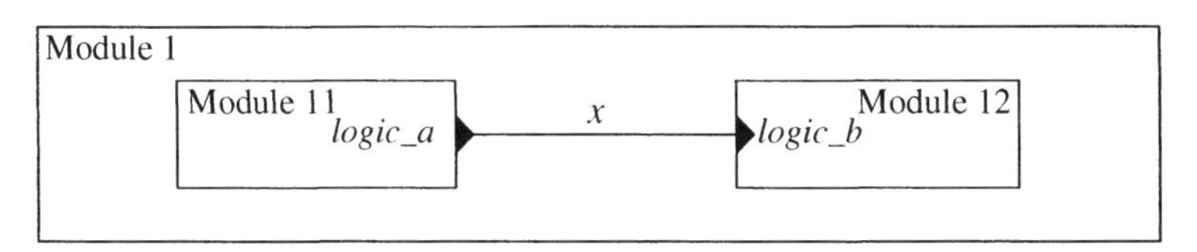

One declares the desired resolution of compatible disciplines using *connect / resolveto* statements:

```
connectrules MyRules;
    connect logic_a, logic_b resolveto logic;
endconnectrules
```

The *connect* statement is used inside of a special block beginning with the keyword *connectrules*. This *connectrules* block defines the discipline resolution and connect module insertion rules for the design. It must be placed outside of the scope of any module. The name of the connect rules, in our case *MyRules*, is unique in the design. Several connect rule definitions with different names can be present in the design and the user uses the names to select which set is used for the simulation. The above connectrules example specifies that a net without a discipline declaration that connects to ports with *logic_a* and *logic_b* disciplines should resolve to the *logic* discipline.

In the analog domain, a design containing interconnections between signal flow and conservative ports is a typical situation where a *resolveto* declaration is needed. Listing 12 on page 119 shows an ADC module using the signal flow discipline *voltage* for the input port *in*. If this input connects on the next higher level of the design hierarchy to an *electrical* port, like a resistor pin, a *resolveto* statement is needed to specify the proper resolved discipline for the net. The statement

```
connect electrical, voltage resolveto electrical;
```

in the *connectrules* block defines that an undeclared net that connects *electrical* and *voltage* pins should take the discipline *electrical.*

3. If one of the lower level nets is analog (continuous-time domain) the higher level net becomes analog too. This means analog wins over digital. It also implies that a connect module will be needed to resolve the incompatibility between the now analog net and any digital (discrete time) ports that connect to it.

With this algorithm the disciplines are resolved upward through the hierarchy, as shown in Figure 5. The structural blocks *block1* and *block2* as well as the behavioral block *d1* are instantiated in the top level. Looking down the hierarchy, *block11* and *d2* are instantiated within *block1*, and *block11* contains *d3* and *d4*. The undeclared net *a* connects the *logic_a* discipline port of *d3* with the *logic_b* discipline port of *d4*. There is no connection to a continuous discipline port at this lowest level of the design hierarchy and so the net *a* will inherit a discrete discipline. By *connectrules MyRules*, *logic_a* and *logic_b* resolve to the *logic* discipline, and so net *a* takes the *logic* discipline (by rule 2). At the next level of the design hierarchy in *block1* with port *a* resolved to the *logic* discipline and with the port of *d2* being *logic*, the undeclared net *b* also inherits the discipline *logic* (by rule 1).

The undeclared net *e* in the top level connects port *b* of *block1* with a *logic* port of *d1* and port *d* of *block2*. The discipline of the undeclared port *d* is not resolved yet. The discipline resolution algorithm must resolve this first before it can determine the discipline of *e*. Since disciplines resolve upward, we must again begin at the lowest level of the design hierarchy. Within *block21* the net *c* connects an *electrical* port of *a1* with a *logic* port of *d6*. The continuous discipline wins over the discrete one (by rule 3) and so net *c* inherits the *electrical* discipline. One level above the now *electrical* port *c* of *block21* is connected to a *logic* port of *d5*. Thus net *d* becomes an *electrical* net (by rule 3). At this point the discipline at the top level can be resolved. Because port *d* of

FIGURE 5 *Basic discipline resolution.*

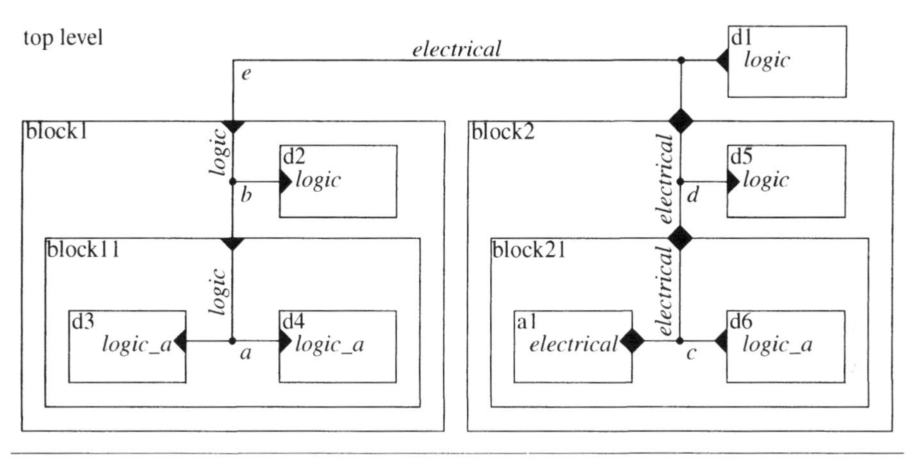

block2 is *electrical* the top level net *e* becomes *electrical* too (by rule 3). All undeclared disciplines are resolved at this point.

Consider another example: the PLL circuit shown in Figure 1 on page 113. Assume that all nets at the top level are undeclared and further assume that the divider (*FD*) and the phase/frequency comparator (*PFD*) are digital models while the charge pump (*CP*), the loop filter (*LF*) and the VCO are analog. The discipline resolution would be simple and straight forward. The net *err* becomes analog because it connects only to analog ports, and net *fb* would become digital because it connects digital ports only. The nets *up*, *dwn* and *out* connect analog and digital ports. Because the analog discipline wins over the digital, they inherit the analog discipline.

4.2.2 Detailed Mixed Signal Discipline Resolution

With the detailed method of discipline resolution every net that in any way connects to an analog port becomes analog. Thus, the basic rule of this method is that analog nets first pass their discipline upwards and then these disciplines propagate downward through the hierarchy to the nets with undeclared disciplines. After this general rule is applied the still unresolved nets inherit the digital discipline from ports that connect from lower in the hierarchy using the basic method of discipline resolution.

Usually detailed discipline resolution results in a larger number of undeclared nets becoming analog as compared to the basic method.

Applying the detailed discipline resolution algorithm to the design in Figure 5 results in the resolved disciplines shown in Figure 6. All shown undeclared wires resolve to the discipline *electrical* because they all connect through the design hierarchy to the *electrical* port of the block *a1*.

FIGURE 6 *Detailed discipline resolution.*

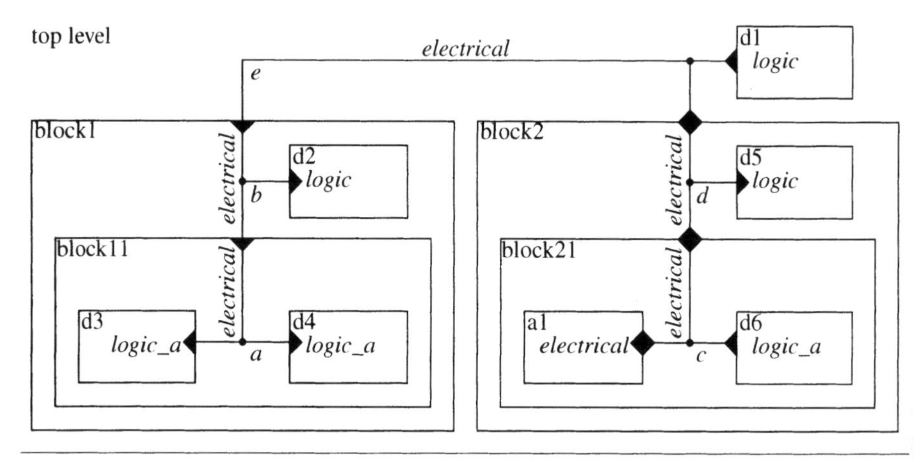

If applied to the PLL example of Figure 1 the detailed discipline resolution result would be identical to the default discipline resolution mechanism because undeclared nets are not connected to ports of undeclared discipline nets from the lower levels of the hierarchy.

After the mixed signal discipline resolution has finished, every net in the design has an assigned discipline, and so each net is associated with a domain.

4.3 Automatic Connect Module Insertion

Connect modules are needed in all cases where a continuous analog discipline net connects to a discrete digital discipline port or where a discrete digital discipline net connects to a continuous discipline port. They are automatically inserted if the proper connect rules are specified. In the example described earlier using the basic discipline resolution algorithm, the insertion points for connect modules are *A*, *B*, *C* and *D* as shown in Figure 7.

The insertion of the connect modules is handled after the discipline resolution is finished and before the simulation starts, typically during the design linking or elaboration phase. At this time Verilog-AMS connect modules have to exist in the simulation library and have to be accessible. They are inserted by name. The automatic insertion

FIGURE 7 *Connect module insertion points.*

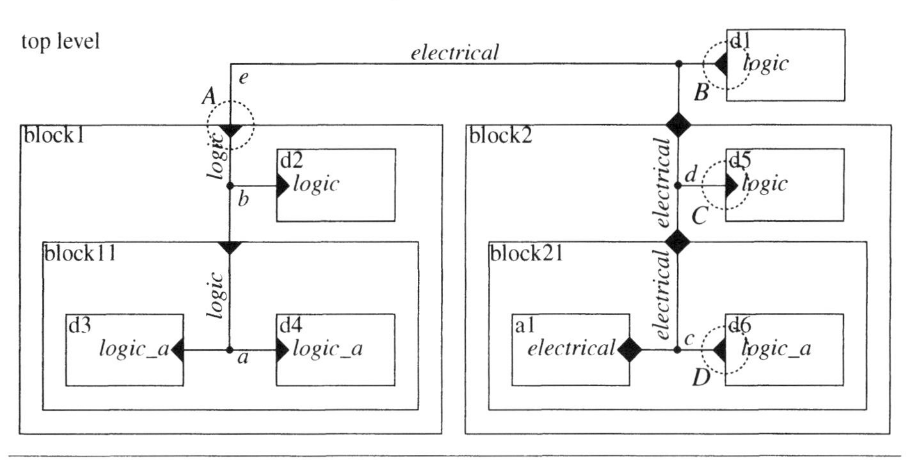

is controlled by the Verilog-AMS *connectrules* block, already introduced in conjunction with the *resolveto* statement. The *connectrules* specification

```
connectrules MyRules;
    connect a2d input electrical, output logic;
    connect d2a input logic, output electrical;
endconnectrules
```

specifies that the connect module with the name *a2d* is used at insertion points that are in need of an *electrical*-to-*logic* discipline transformation. This is the case at input ports of *logic* discipline that are driven by an *electrical* net. In the example, *a2d* connect modules will be inserted at *A* and *C*. The *electrical* input of the *a2d* module is connected to the *electrical* net, and the *logic* output of the *a2d* module to the *logic* input port pin as shown in Figure 8. In the other direction the module with the name *d2a* is inserted where *logic*-to-*electrical* discipline transformation is necessary. In Figure 7 this is required at the points *B* and *D*.

For the PLL example of Figure 1 an *u2d* element is inserted between the *electrical* net *out* and the divider input port with *d2a* modules being inserted between both *logic* outputs of the phase comparator and the *electrical* nets *up* and *dwn*.

Verilog-AMS provides two different modes of connect module insertion: merged and split. Merged insertion is the default. In this mode all of the connect modules of the same kind at the same net and at the same level of the design hierarchy are merged into one. In the split mode separate connect modules are inserted; one for each of the

FIGURE 8 *Connect module insertion.*

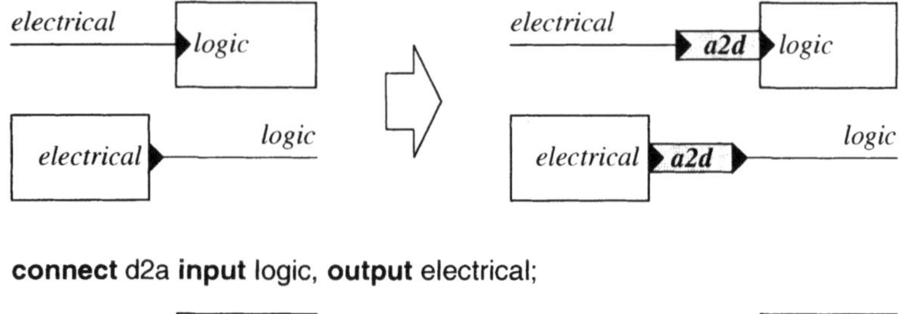

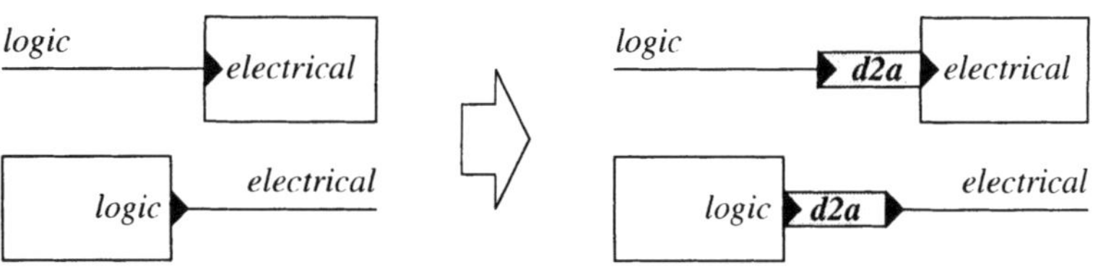

port pins that connects to the net. The *split* and *merged* keywords in the *connect* statement specify which insertion mode is used. The definition

```
connectrules MyRules;
    connect a2d split input electrical, output logic;
    connect d2a merged input logic, output electrical;
endconnectrules;
```

specifies the connect module with the name *a2d* use split insertion and the *d2a* modules use merged insertion.

Setting the insertion mode to split together with enabling detailed discipline resolution leads to the highest number of connect modules being inserted into the design, whereas the default mode for both (merged insertion and basic discipline resolution) result in the fewest insertions. Because increasing numbers of interface components generally slow the simulation, these default settings usually lead to the fastest simulation.

For our PLL example whether the connect modules are merged or split makes no difference. But it makes a difference for the example in Figure 6. Using detailed discipline resolution the *logic* inputs of *d3* and *d4* are connected to the same *electrical* net *a* within *block11*. With the split insertion mode 2 separate connect modules between *a* and both *logic* inputs would be inserted. In the default merged mode the discipline conversion is handled by one common module.

4.4 Modeling Connect Modules

The basic task of connect modules is the transformation of continuous information into discrete, or visa versa. Consider an *electrical*-to-*logic* connect module: it has to convert the continuous changing voltage value at its *electrical* input into an appropriate logic state at its *logic* output. But there is more than just converting the voltage value. The mixed signal simulation delivers the right results, comparable with the real circuit, if the behavior of the connect module input is very close to the behavior of the digital gate in the real circuit. To ensure the simulation mimics reality as closely as possible, impedance, nonlinearity, supply voltage dependency and such have to be considered. Also, at the digital output of the module, simply switching the logic state when the input voltage crosses a defined threshold value might not be sufficient to match the real circuit behavior. For instance, it might be important to model the influence of the rise and fall times of the analog input voltage, hysteresis effects, or the dependency on the digital supply voltage. In general we can state that the same compromise must be made between model accuracy and simulation speed as with any other model.

The full Verilog-AMS capabilities are available when creating these connect modules. Every regular module that has two single ports, one in the analog domain and the other one in the digital domain could be used as a connect module. However, a model that is used for automatic insertion needs to have the *connectmodule* keyword instead of the *module* keyword in the module header.

```
connectmodule a2d (d, a);
    input a;
    output d;
    electrical a;
    logic d;
```

With connection modules, either one port must be an *input* and the other an *output*, or both must be *inout*. The ports may be given in either order. We have chosen to follow the Verilog convention of putting the output port first.

4.4.1 Analog to Digital Connect Modules

A basic analog-to-digital connect module is shown in Listing 16. It monitors its analog input and produces a digital 1 at its output if the input goes above the upper threshold, *vh*. It produces a digital 0 at its output if the input goes below the lower threshold, *vl*. The example demonstrates the use of the *above* analog event function called inside of the digital *always* block to detect the threshold crossing (5§6.8.3p206). The analog equation

```
I(in) <+ c*ddt(V(in));
```

LISTING 16 *Basic electrical to logic connect module.*

```
`include "disciplines.vams"

connectmodule a2d (out, in);
    parameter real vh = 2.7;   // minimum voltage of a logic 1 (V)
    parameter real vl = 0.5;   // maximum voltage of a logic 0 (V)
    parameter real c = 100f;   // input capacitance (F)
    input in;
    output out;
    electrical in;
    reg out;
    logic out;

    // when analog rises above the high threshold, digital becomes 1
    always @(above(V(in) – vh))
        out = 1'b1;

    // when analog falls below the low threshold, digital becomes 0
    always @(above(vl – V(in)))
        out = 1'b0;

    analog
        I(in) <+ c*ddt(V(in));
endmodule
```

models the input capacitance of the connected digital gate (3§1.1p39). The connect module parameters *vl*, *vh* and *c* can be overwritten when specifying this module in the *connectrules* block:

```
connect a2d #(.vl(1.2), .vh(2.2), .c(1p));
```

This is an easy way to make connect modules fit the process parameters. The above connect statement does not specify the input and output disciplines as in the previous example. The discipline specification is optional. In this case the connect module *a2d* is used for all mixed signal boundaries that match the input and output declaration of the connect module header. The input *in* is declared as *electrical* and output *out* as *logic* discipline. Thus it will be used automatically for the *electrical* to *logic* boundaries.

If the simulated design contains different types of logic, for instance 3 and 5 volt logic, a good way for the user to control the insertion of connect modules is by assigning different digital disciplines to logic operating at different supply voltages. However, there is no need to have separate connect modules available for insertion at the boundaries of these different logic types to *electrical*. The same connect module can be used for both, with parameters used to account for the different supply voltages.

```
connect a2d input electrical, output logic5;
connect a2d #(.vl(0.9), .vh(2.1)) input electrical, output logic3;
```

The first statement defines that the module with the name *a2d* is used at *electrical* to *logic5* boundaries using the default parameter values declared in the module for the thresholds. The second connect statement defines that the same module is automatically inserted at the *electrical* to *logic3* boundaries with different values for *vl* and *vh*.

The connect module example in Listing 17 contains an important enhancement. It takes the input voltage range between the two voltage thresholds into account. If the input voltage remains in the range between the low threshold *vl* plus a cushion of *dv* and the high threshold *vh* minus a cushion of *dv* longer than a certain delay time *dt* the module outputs a logic *x* state. This is illustrated in Figure 9. This model is appropriate if there is a risk that the rise or fall times at the inputs of the logic gates might not be fast enough to ensure clean switching.

FIGURE 9 *Output characteristics of enhanced electrical to logic connect module.*

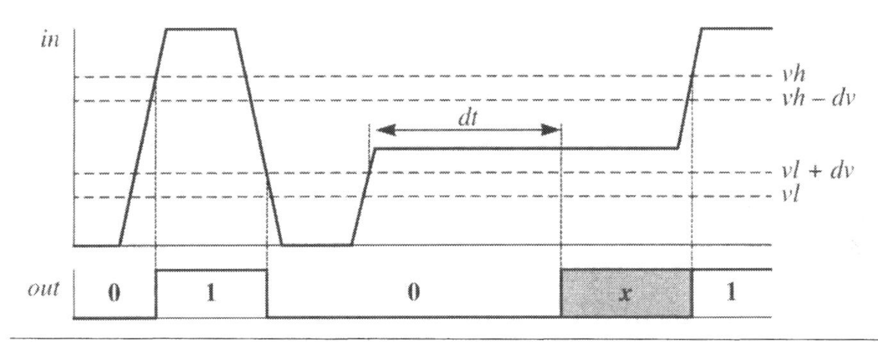

If the input voltage rises above the $vlx = vl + dv$ threshold or if it falls below the $vhx = vh - dv$ threshold the register *inXregion* is set to 1 with the *always* blocks

```
always @(above(vhx – V(in))) inXregion = 1;
always @(above(V(in) – vlx)) inXregion = 1;
```

which trigger another *always* block

```
always @(posedge inXregion) begin : XRegionDelay
```

This is a named block (5§7.2.1p210). The block name follows the colon after the *begin* keyword. After a wait time equal to *dt* the output turns to *x*. However, if the input voltage falls below the *vl* threshold or rises above the *vh* threshold in the mean time, then *inXregion* is reset to 0 within another set of *always* blocks, which triggers the *always* block

LISTING 17 *Enhanced electrical to logic connect module.*

```
`timescale 1ns / 1ps
`include "disciplines.vams"

connectmodule a2d (out, in);
    parameter real vh = 2.7;        // minimum voltage of a logic 1 (V)
    parameter real vl = 0.5;        // maximum voltage of a logic 0 (V)
    parameter real c = 100f;        // input capacitance (F)
    parameter real dt = 1n;         // time in x region before x is produced (s)
    parameter real dv = 0.2;        // voltage between threshold and x region (V)
    input in;
    output out;
    electrical in;
    reg out;
    logic out;

    parameter real vlx = vl + dv;
    parameter real vhx = vh – dv;
    reg inXregion;

    initial inXregion = 0;

    always @(above(V(in) – vh)) begin
        out = 1'b1;
        inXregion = 0;
    end

    always @(above(vl – V(in))) begin
        out = 1'b0;
        inXregion = 0;
    end

    always @(above(vhx – V(in))) inXregion = 1;
    always @(above(V(in) – vlx)) inXregion = 1;

    always @(posedge inXregion) begin : XRegionDelay
        #(dt/1.0n) // calculate how many time units are equal to the dt time
        out = 1'bx;
        inXregion = 0;
    end
    always @(negedge inXregion) disable XRegionDelay;

    analog
        I(in) <+ c*ddt(V(in));
endmodule
```

```
always @(negedge inXregion) disable XRegionDelay;
```

that disables the named block *XRegionDelay* (5§7.2.1p210). Thus, in this case, *out* would not turn to *x* after the *dt* time. That means, if the input voltage crosses the forbidden input range within a time shorter than *dt*, the output switches between the 0 and the 1 states directly. But if the transition takes longer than *dt*, an *x* state would be produced at the module output in between the 0 and the 1.

4.4.2 Digital to Analog Connect Modules

A simple digital-to-analog connect module is constructed as shown in Listing 18. In this model the digital input controls both the output voltage and resistance of the converter. In this way, the *z* state can be accurately modeled. The behavior of the module is simple. The input signal is monitored, and the value of the discrete real variables *v* and *r* are updated to the appropriate values initially and whenever the input changes. The values of *v* and *r* are then used in the analog process to model the analog output. Both *v* and *r* are passed through transition functions to add a finite transition time whenever their values change (5§4.6.4p180).

4.4.3 Bidirectional Connect Modules

Especially in large SOC designs bidirectional mixed-domain interfaces are required. There are blocks, represented as analog behavior or structure, connected to bidirectional communication data busses. That is, the analog block receives information and is also able to send information over the same connection. On the digital side there could be several blocks with *inout* connections at this bus. Each bus wire needs an *inout* connection module. The difficulty is that the connection module for automatic insertion during the elaboration phase can only have one pin on the digital side. That is, the connect module needs to be able to read and write at the same time via one single port to ensure true bidirectional behavior.

The importance of this can be seen by considering an application of the digital-to-analog connect module of Listing 18. With this connect module there is no feedback from the analog side back to the digital side: the connection is unidirectional. This can be problematic if the signal level on the analog side of the connect module never achieves the desired value. For example, if the connect module receives as input a logical 1, but because of loading effects, the output voltage remains near 0 V, then the analog and digital representations of the same signal are inconsistent. This inconsistency would manifest itself as errors in the case where part of the system is monitoring the signal value on the analog side, and where another part is monitoring the signal on the digital side, as is the case shown in Figure 10.

LISTING 18 *Simple digital-to-analog connect module.*

```
`include "disciplines.vams"
`timescale 1ns / 10ps

connectmodule d2a (out, in);
    parameter real v0 = 0.0;                    // output voltage for a logic 0 (V)
    parameter real v1 = 5.0;                    // output voltage for a logic 1 (V)
    parameter real vx = 2.5;                    // output voltage for a logic x (V)
    parameter real vz = 5.0;                    // output voltage for a logic z (V)
    parameter real r0 = 1k from (0:inf);        // output resistance for a logic 0 (Ohms)
    parameter real r1 = 1k from (0:inf);        // output resistance for a logic 1 (Ohms)
    parameter real rx = 100 from (0:inf);       // output resistance for a logic x (Ohms)
    parameter real rz = 1M from (0:inf);        // output resistance for a logic z (Ohms)
    parameter real tr=1n from [0:inf);          // rise time (s)
    parameter real tf=1n from [0:inf);          // fall time (s)
    input in;
    output out;
    logic in;
    electrical out;
    real v, r;

    assign in = in; // connect digital drivers and receivers

    initial begin
        case(in)
            1'b0: begin  v = v0;  r = r0;  end
            1'b1: begin  v = v1;  r = r1;  end
            1'bx: begin  v = vx;  r = rx;  end
            1'bz: begin  v = vz;  r = rz;  end
        endcase
    end

    always @in begin
        case(in)
            1'b0: begin  v = v0;  r = r0;  end
            1'b1: begin  v = v1;  r = r1;  end
            1'bx: begin  v = vx;  r = rx;  end
            1'bz: begin  v = vz;  r = rz;  end
        endcase
    end

    analog
        V(out) <+ transition(v, 0, tr, tf) + transition(r, 0, tr, tf)*I(out);
endmodule
```

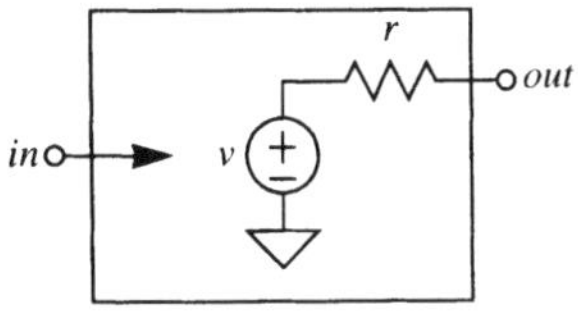

in — v — r — out

in	*v*	*r*
0	0 V	1 kΩ
1	5 V	1 kΩ
x	2.5 V	100 Ω
z	5 V	1 MΩ

FIGURE 10 *Bidirectional mixed signal connection.*

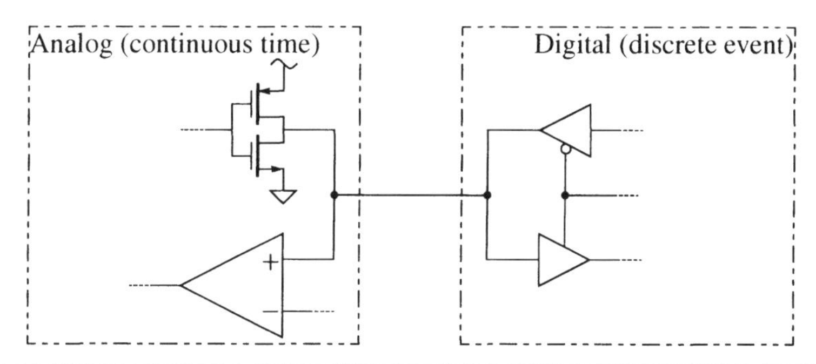

To resolve the inconsistency, the connect module must pass information in both directions. How can we feed information about the analog solution back to the digital side and use that information to change the value observed on the digital net? The Verilog-AMS language supports this with the ***driver/receiver segregation*** feature. Digital drivers and digital receivers are separated on digital wires connected to connect modules as shown in Figure 11. Effectively, the net is split with all the drivers attached to one side and all the receivers attached to the other. The connect module then acts as the bridge. As such, the connect module must drive the net, even if the digital port is an input port, and in doing so it is driving the receivers. In the unidirectional *d2a* module shown in Listing 18 we simply pass the resolved value of the driven portion of the net on to the portion being monitored by the receivers with

```
assign in = in;
```

A bidirectional connect module instead combines the resolved value of the driven portion of the digital net with the resolved signal level on the analog side of the connect module to determine the value that is passed on to the receivers monitoring the digital net. The bidirectional connect module of Listing 19 uses this feature.

From Digital to Analog. Assuming for the moment there is only one logic block connected to the wire, the connection module inserted for this wire reads what the digital block drives on the net. The state could be an active 0 or 1, it could be undefined x, or it could be undriven, z. This information must be translated to analog and applied with the right driver strength at the analog output node. The 0 and 1 states should apply an appropriate voltage v with a relatively small output resistance r to the analog output node and z should apply a voltage with a large output resistance (see Figure 12). The undefined x state is handled in a special way. If driven to the *logic* input of the connection module, an x is directly applied to the digital receivers. In this case the voltage source and resistance applied to the analog output node will be set by user-defined

FIGURE 11 *A digital port on a bidirectional connect module, before and after driver/receiver segregation.*

Bidirectional connect module (as specified)

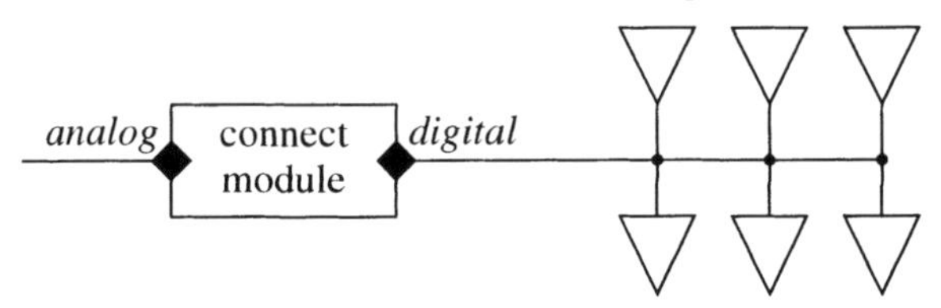

Bidirectional connect module (with drivers and receivers segregated)

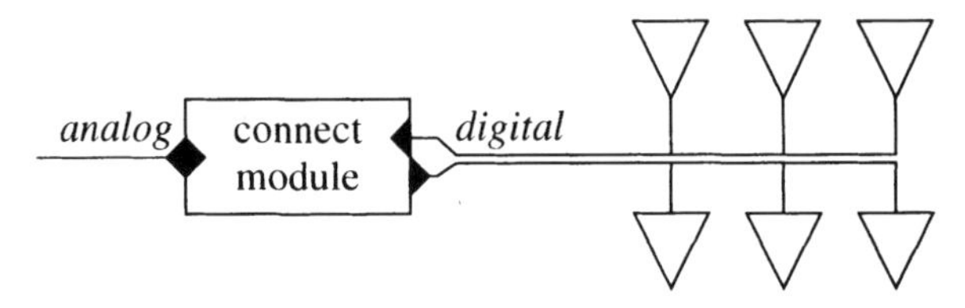

parameters. Because there is no undefined state in the continuous analog domain, it is the choice of the simulation user to decide which output voltage and resistance should be used in this case. The examples in this book use an intermediate value for the voltage and a very small output resistance in an attempt to force the analog circuitry to see an indeterminate voltage level independent of loading.

FIGURE 12 *Bidirectional connect module.*

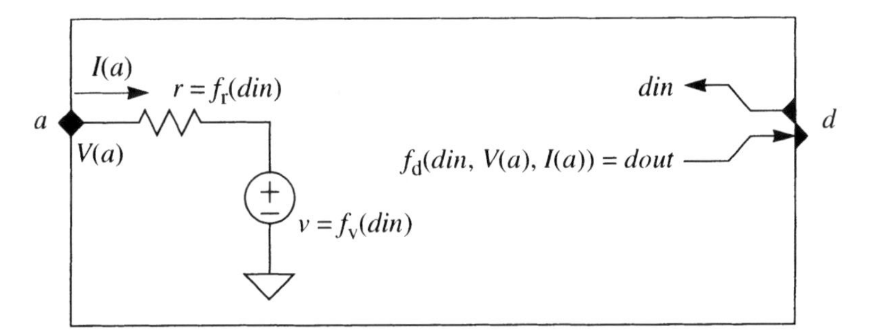

From Analog to Digital. The analog solver of the mixed signal simulator resolves operating conditions of the analog output node of the connection element in the context of the larger analog circuit to which it is connected. The resulting port voltage

V(*a*) and port current *I*(*a*) on the analog side of the wire are used when determining the logical value at this port. The resolved logical value of the analog port is then used when determining *dout*, the value that is fed back to the receivers being driven by the digital port of the connection module. If the voltage at the analog connection is below the 0-threshold or above the 1-threshold, the 0 or 1 state is assigned to the digital *inout* port accordingly. If *v* leaves the 0 or 1 range perhaps after an edge of the digital input signal was detected, *dout* stays unchanged for the first moment. If the voltage remains in the *x*-range for longer than a preset time limit *xdelay*, the output *dout* will change to *x* and stay *x* until the voltage moves back into the 0 or 1 range. This way transitions that are too slow are detected. The *z*-condition at the analog connection module node is detected by using the port current. Detection is possible only if the designated *z*-voltage recognition level is within the *x*-range. If this is the case when a *z* is detected at the digital input and the absolute current into the analog connect module port is less than a certain limit, then *dout* will be set to *z*.

The statement

```
assign d = dout;
```

in Listing 19 applies the digital output state to the receivers connected to the *logic inout* port of the connect module. Cross statements are used for a precise detection of the voltage and the current threshold crossings related to the analog connect module port. The signals *vstate* and *istate* are used to monitor the analog output voltage and current (see Tables 4 and 5). The signal *inXrange* is used to show the analog voltage being in the *x*-range but for not longer than the time limit parameter *xdelay*. When the connect module reads an undefined *x* on the *logic* side it also applies this *x* state back to the receivers. Table 3 shows how *dout,* which is fed back to the receivers, depends on the digital input *d* as well as the *istate* and *vstate* signals.

TABLE 3 *Function table for receiver information.*

d	**vstate**	**istate**	**dout**
x	*d*	*d*	*x*
z	1	*d*	1
z	2	0	*z*
z	2	1	*x*
z	3	*d*	0
1	1	*d*	1
1	2	*d*	*x*

TABLE 3 *Function table for receiver information.*

d	vstate	istate	dout
1	3	d	x
0	1	d	x
0	2	d	x
0	3	d	0

TABLE 4 *Function table for vstate.*

Condition	vstate
$V(a) >$ vh	1
vh $\geq V(a) \geq$ vl	2
vl $> V(a)$	3

TABLE 5 *Function table for istate.*

Condition	istate
maxzi $\geq \|I(a)\|$	0
$\|I(a)\| >$ maxzi	1

The above example contains a minimum set of bidirectional connect module features. Enhancing this model means adding features and bringing the model closer to reality. The user should be aware that adding features may also result in lowering the simulation speed. It is usually helpful to have a variety of connect modules with different detail levels available. That way the connection module could be chosen to provide the best trade-off between accuracy and speed.

One of the first enhancements you might think to add is power supply dependency for the build-in thresholds and the output voltage levels. This could be done easily by binding the connect module parameters to a global power supply parameter. This way a static power supply dependency is realized. If the simulation of the design requires dynamic power supply dependency in the connect module, it can be implemented by referencing or connecting to global power supply nodes from inside the connect module using hierarchical names (5§9.4p230). Effects like power on/off behavior or tran-

LISTING 19 *Bidirectional connect module.*

```
`include "disciplines.vams"
`timescale 1ns / 10ps

connectmodule bidir(a, d);
    parameter real v0 = 0.0;                  // output voltage for a logic 0 (V)
    parameter real v1 = 5.0;                  // output voltage for a logic 1 (V)
    parameter real vx = 2.5;                  // output voltage for a logic x (V)
    parameter real vz = 5.0;                  // output voltage for a logic z (V)
    parameter real r0 = 1k from (0:inf);      // output resistance for a logic 0 (Ohms)
    parameter real r1 = 1k from (0:inf);      // output resistance for a logic 1 (Ohms)
    parameter real rx = 100 from (0:inf);     // output resistance for a logic x (Ohms)
    parameter real rz = 1M from (0:inf);      // output resistance for a logic z (Ohms)
    parameter real tr=1n from [0:inf);        // rise time (s)
    parameter real tf=1n from [0:inf);        // fall time (s)
    parameter real vl = 0.5;                  // maximum voltage of a logic 0 (V)
    parameter real vh = 2.7;                  // minimum voltage of a logic 1 (V)
    parameter real maxzi = 0.1u;              // absolute max current allowed in HiZ state
    parameter real xdelay = 1;                // time in the Xrange before dout gets x
    parameter real zdelay = 0.5 * xdelay;     // max time of current > maxzi before
                                              // going out of z state
    inout a, d;
    electrical a;
    logic d;
    reg dout;
    logic dout;
    real v, r;
    integer vstate;
    reg istate, inXrange, outOfZcurrent;

    assign d = dout;

    initial begin
        dout = 1'bz;        // set the digital output to z until analog voltage is resolved
        vstate = 2; istate = 0; inXrange = 0; outOfZcurrent = 0;
        case (d)
            1'b0: begin  v = v0;  r = r0;               end
            1'b1: begin  v = v1;  r = r1;               end
            1'bx: begin  v = vx;  r = rx;  dout=1'bx;   end
            1'bz: begin  v = vz;  r = rz;               end
        endcase
    end
```

Continued on next page.

LISTING 19 *Bidirectional connect module.*

Continued from previous page.

```
always @d begin
    case(d)
        1'b0: begin  v = v0;  r = r0;                end
        1'b1: begin  v = v1;  r = r1;                end
        1'bx: begin  v = vx;  r = rx;   dout=1'bx;   end
        1'bz: begin  v = vz;  r = rz;                end
    endcase
end

always @(above(V(a) – vh)) begin vstate = 1; inXrange = 0; end
always @(above(vh – V(a))) inXrange = 1;
always @(above(V(a) – vl, 1)) inXrange = 1;
always @(above(vl – V(a))) begin vstate = 3; inXrange = 0; end

always @(posedge inXrange) begin : XRangeDelay
    #xdelay
    vstate = 2;
    inXrange = 0;
end
always @(negedge inXrange) disable XRangeDelay;

always @(posedge outOfZcurrent) begin : outOfZCurrentDelay
    #zdelay
    istate=1;
    outOfZcurrent=0;
end
always @(negedge outOfZcurrent) disable outOfZCurrentDelay;

always @(above(abs(I(a)) – maxzi)) outOfZcurrent = 1;
always @(above(maxzi – abs(I(a)))) begin
    istate=0;
    outOfZcurrent=0;
end

always @(vstate or istate or d) begin
    case(vstate)
        1: dout = (d === 1'bx) ? 1'bx : 1'b1;
        2: dout = ((istate === 1'b0) & (d===1'bz)) ? 1'bz : 1'bx;
        3: dout = (d === 1'bx) ? 1'bx : 1'b0;
    endcase
end
```

Continued on the next page.

LISTING 19 *Bidirectional connect module.*

Continued from the previous page.

```
    analog begin
        V(a) <+ transition(v,0,tr,tf) + transition(r,0,tr,tf) * I(a);
    end
endmodule
```

sient power supply noise become visible. The modeling must be done in the analog block if the remote voltage varies in continuous time. One might use code like the following to model this effect.

```
    analog begin
        case (d)
            1'b0: v = V(top.vdd);
            1'b1: v = V(top.gnd);
            1'bx: v = vx;
            1'bz: v = vz;
        endcase

        V(a) <+ v + r * I(a);
    end
```

This assumes that *top.vdd* and *top.gnd* are the global supply and ground nodes, that *vx* and *vz* are parameters, and that *d* and *r* are set appropriately earlier in the connect module.

Another possible enhancement is to make the characteristics of the analog port dependent on the number and the states of the drivers at the digital side. Such a model is illustrated in Figure 13 and given in Listing 20.

FIGURE 13 *A digital port on a bidirectional connect module with driver/receiver segregation and access to the values of individual drivers.*

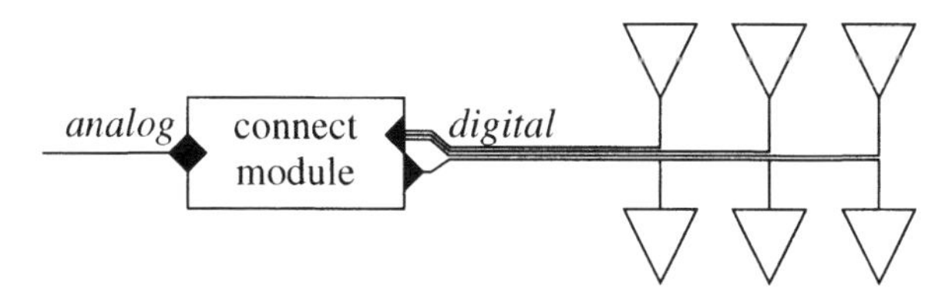

Information about the digital drivers on a digital net can be accessed from inside of the connect module. The driver access function

$driver_count(*wire name***)**

returns the number of drivers connected to the wire. It is generally best to ask for the number of drivers on a wire in the initial block, because the number of drivers will not change during the simulation. The function

$driver_state(*wire name, driver index***)**

returns the status of the specified driver, and the function

driver_update(*wire name***)**

returns an event whenever at least one of the connected drivers on this net changes its state.

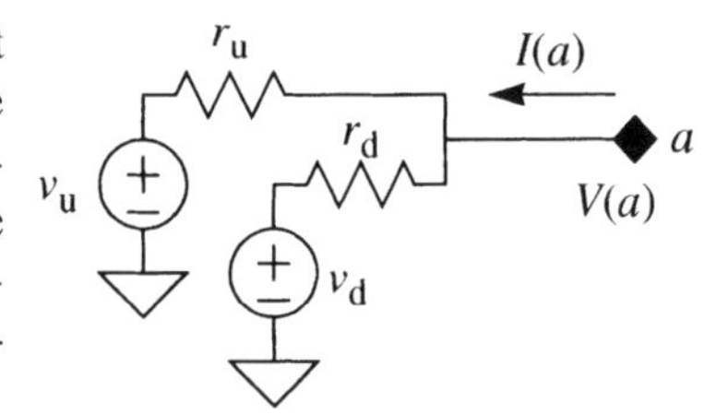

Conceptually, the analog port of the connect module is driven using the circuit shown on the right. v_u and v_d represent the pull-up and pull-down voltages. They remain fixed. r_u and r_d are the pull-up and pull-down resistances, their values are changed depending on the number of digital drivers and their state.

The values of r_u and r_d are

$$r_u = \frac{r_{u\,on}}{n_1 + n_x} \| \frac{r_{u\,off}}{n_0 + n_z} \text{ and} \tag{1}$$

$$r_d = \frac{r_{d\,on}}{n_0 + n_x} \| \frac{r_{d\,off}}{n_1 + n_z}, \tag{2}$$

where $r_{u\,on}$, $r_{u\,off}$, $r_{d\,on}$ and $r_{d\,off}$ are parameters, n_1, n_0, n_x and n_z are the number of drivers whose state is 0, 1, x and z respectively, and $\|$ represents the parallel combination of resistor values (if $r_p = r_1 \| r_2$ then $r_p = 1/(1/r_1 + 1/r_2)$).

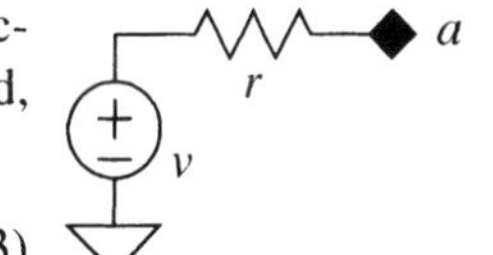

From an implementation perspective, using this particular structure is inefficient because it requires two internal nodes. Instead, the Thevenin equivalent (right) is used where

$$r = r_u \| r_d \text{ and} \tag{3}$$

$$v = \frac{r_d v_u + r_u v_d}{r_u + r_d}. \tag{4}$$

It may seem like this implementation also uses an internal node, but Verilog-A/MS allows the two branches to be combined into one, eliminating the extra node. Often times even more efficiency can be gained by switching to a Norton equivalent, but for simplicity we will stay with the Thevenin equivalent.

As before, the port voltage and port current are used to determine the logical state applied to the digital receivers.

Verilog-AMS provides additional *$driver_...* functions besides the ones described here that allow you to access additional information about the drivers, such as their type, their strength, and information about known upcoming transitions. These functions allow you to build even more sophisticated connect modules. Information about these functions can be found in the Verilog-AMS LRM [28].

What's Next

This chapter introduced the discrete-event behavioral modeling for both digital and mixed-signal systems, and it also covered mixed-signal structural modeling. At this point, all of the important concepts of Verilog-AMS have been introduced. However, many details have been left out to simplify the presentation. These details can be found in the next chapter.

LISTING 20 *Bidirectional connect module with driver access.*

```
`include "disciplines.vams"
`timescale 1ns / 10ps

connectmodule bidir(a, d);
    parameter real v0 = 0.0;                  // output voltage for a logic 0 (V)
    parameter real v1 = 5.0;                  // output voltage for a logic 1 (V)
    parameter real ruon = 1k from (0:inf);    // pull-up resistance when 1 or x (Ω)
    parameter real ruoff = 1M from (0:inf);   // pull-up resistance when 0 or z (Ω)
    parameter real rdon = 1k from (0:inf);    // pull-down resistance when 1 or x (Ω)
    parameter real rdoff = 1M from (0:inf);   // pull-down resistance when 0 or z (Ω)
    parameter real tr=1n from [0:inf);        // rise time (s)
    parameter real tf=1n from [0:inf);        // fall time (s)
    parameter real vl = 0.5;                  // maximum voltage of a logic 0 (V)
    parameter real vh = 2.7;                  // minimum voltage of a logic 1 (V)
    parameter real maxzi = 0.1u;              // absolute max current allowed in HiZ state
    parameter real xdelay = 1;                // time in the Xrange before dout gets x
    parameter real zdelay = 0.5 * xdelay;     // max time of current > maxzi before
                                              // going out of z state
    inout a, d;
    electrical a;
    logic d;

    parameter real guon = 1.0/ruon;       // these are local constants
    parameter real guoff = 1.0/ruoff;     // they should not be overridden by the user
    parameter real gdon = 1.0/rdon;
    parameter real gdoff = 1.0/rdoff;
    real v, r, gu, gd;
    integer vstate;
    reg istate, inXrange, outOfZcurrent;
    integer DrCount, i;
    integer XCount, ZCount, LCount, HCount;
    reg dout;
    logic dout;

    assign d = dout;

    initial begin
        dout = 1'bz; // set the digital output to z until analog voltage is resolved
        vstate = 2; istate = 0; inXrange = 0; outOfZcurrent = 0;
        DrCount = $driver_count(d);
        XCount = 0; ZCount = 0; LCount = 0; HCount = 0;
```

Continued on the next page.

LISTING 20 *Bidirectional connect module with driver access.*

Continued from the previous page.

```
        for (i = 0; i < DrCount; i = i + 1)
            case($driver_state(d,i))
                1'b0: LCount = LCount+1;
                1'b1: HCount = HCount+1;
                1'bx: begin XCount = XCount+1; dout=1'bx; end
                1'bz: ZCount = ZCount+1;
            endcase
        gu = (XCount+HCount)*guon + (ZCount+LCount)*guoff;
        gd = (XCount+LCount)*gdon + (ZCount+HCount)*gdoff;
        r = 1.0/(gu + gd);
        v = (gu*v1 + gd*v0)/(gu + gd);
    end

    always @(driver_update(d)) begin
        XCount = 0; ZCount = 0; LCount = 0; HCount =0;
        for (i = 0; i < DrCount; i = i + 1)
            case($driver_state(d, i))
                1'b0: LCount = LCount+1;
                1'b1: HCount = HCount+1;
                1'bx: begin XCount = XCount+1; dout = 1'bx; end
                1'bz: ZCount = ZCount+1;
            endcase
        gu = (XCount+HCount)*guon + (ZCount+LCount)*guoff;
        gd = (XCount+LCount)*gdon + (ZCount+HCount)*gdoff;
        r = 1.0/(gu + gd);
        v = (gu*v1 + gd*v0)/(gu + gd);
    end

    always @(above(V(a) – vh)) begin vstate = 1; inXrange = 0; end
    always @(above(vh – V(a))) inXrange = 1;
    always @(above(V(a) – vl)) inXrange = 1;
    always @(above(vl – V(a))) begin vstate = 3; inXrange = 0; end

    always @(posedge inXrange) begin : XRangeDelay
        #xdelay
        vstate = 2;
        inXrange = 0;
    end
    always @(negedge inXrange) disable XRangeDelay;
```

Continued on the next page.

LISTING 20 *Bidirectional connect module with driver access.*

Continued from the previous page.

```
    always @(posedge outOfZcurrent) begin : outOfZCurrentDelay
        #zdelay
        istate = 1;
        outOfZcurrent = 0;
    end
    always @(negedge outOfZcurrent) disable outOfZCurrentDelay;

    always @(above(abs(I(a)) – maxzi)) outOfZcurrent=1;
    always @(above(maxzi – abs(I(a)))) begin
        istate = 0;
        outOfZcurrent = 0;
    end

    always @(vstate or istate or d) begin
        case(vstate)
            1: dout= (d === 1'bx) ? 1'bx : 1'b1;
            2: dout= ((istate === 1'b0) & (d === 1'bz)) ? 1'bz : 1'bx;
            3: dout= (d === 1'bx) ? 1'bx : 1'b0;
        endcase
    end

    analog begin
        V(a) <+ transition(v,0,tr,tf) + transition(r,0,tr,tf) * I(a);
    end

endmodule
```

5 Language Reference

1 Basics

Verilog-A/MS is a case sensitive language.

Spaces, tabs, and newlines are considered white space and are ignored except when found in strings.

1.1 Comments

Comments are text added to the model for purposes of documentation. They are ignored by the simulator that implements the model.

Single line comments start with // and end at the end of the line.

```
// this is a single line comment
```

Block comments begin with /* and end with */.

```
/*
 * This is a block comment
 */
```

Block comments may not be nested.

1.2 Identifiers

An ***identifier*** is used to give an object a unique name so it can be referenced. An identifier can be any sequence of letters, digits, dollar signs '$', and the underscore characters '_'. The first character of an identifier cannot be a digit or '$'; it can be a letter or an underscore.

Examples: clk, out_p, bus2, n$12

Escaped identifiers start with the backslash character '\' and end with white space (space, tab or newline). They provide a means of including any of the printable ASCII characters in an identifier. Neither the leading back-slash character nor the terminating white space is considered to be part of the identifier. Therefore, an escaped identifier \out is treated the same as a non-escaped identifier out.

Examples: \out+, \/x1/n1, \\x1\n1, \{a,b}, \V(p,n)

1.3 Keywords

Keywords are predefined non-escaped identifiers that are used to define the language constructs. The list of reserved keywords for Verilog-AMS is shown in Table 1. Preceding a keyword with an escape character (\) causes it to be interpreted as an escaped identifier.

TABLE 1 *Reserved words in Verilog-AMS*

above	cross	forever	ln	pulldown	time
abs	ddt	fork	log	rcmos	timer
absdelay	deassign	from	macromodule	real	tran
ac_stim	default	function	max	realtime	tranif0
acos	defparam	generate	medium	reg	tranif1
acosh	disable	genvar	min	release	transition
always	discipline	ground	module	repeat	tri
analog	driver_update	highz0	nand	rnmos	tri0
analysis	edge	highz1	nature	rpmos	tri1
and	else	hypot	negedge	rtran	triand
asin	end	idt	net_resolution	rtranif0	trior
asinh	endcase	idtmod	nmos	rtranif1	trireg
assign	endconnectrules	if	noise_table	scalared	vectored
atan	enddiscipline	ifnone	nor	sin	wait
atan2	endfunction	inf	not	sinh	wand
atanh	endmodule	initial	notif0	slew	weak0
begin	endnature	initial_step	notif1	small	weak1
branch	endprimitive	inout	or	specify	while
buf	endspecify	input	output	specparam	white_noise
bufif0	endtable	integer	parameter	sqrt	wire
bufif1	endtask	join	pmos	strong0	wor
case	event	laplace_nd	posedge	strong1	wreal
casex	exclude	laplace_np	potential	supply0	xnor
casez	exp	laplace_zd	pow	supply1	xor
ceil	final_step	laplace_zp	primitive	table	zi_nd
cmos	flicker_noise	large	pull0	tan	zi_np
connectrules	flow	last_crossing	pull1	tanh	zi_zd
cos	force	limexp	pullup	task	zi_zp
cosh					

1.4 Compiler Directives

The ` character (referred to as a tick, an open quote, or a grave accent) introduces a language construct used to implement compiler directives. The behavior dictated by a compiler directive takes effect as soon as the compiler reads the directive. The directive remains in effect for the rest of the compilation unless a different compiler directive specifies otherwise. A compiler directive in one file can therefore control compilation behavior in multiple description files.

Verilog-AMS supports the following compiler directives.

`default_discipline	`else	`resetall
`default_transition	`endif	`timescale
`define	`ifdef	`undef
	`include	

Defines (*`define*) give a name to a string that can substitute for a string of characters. The name is then referred to as a macro. Any valid identifier, including keywords already in use, can be used as a name. Once defined, the macro is referenced using its name preceded by a tick. Undefines (*`undef*) remove the macro.

Example:

```
`define size 8
electrical [0:`size–1] out;
```

Includes (*`include*) are replaced by the contents of a file. It takes the filename as an argument, which can either be specified with a relative or absolute path to the file. Included files may include other files, etc.

Example: **`include** "disciplines.vams"

Sections of code can be conditionally ignored using the *`ifdef* directive. It takes a macro name as an argument. If the argument is currently undefined, the text that follows is ignored up to a matching `else or `endif and accepted otherwise. If `else is used, then the text between it and the matching `endif is ignored if the argument is defined, and accepted otherwise.

Verilog-AMS supports a predefined macro to allow modules to be written that work with both IEEE 1364-1995 Verilog HDL and Verilog-AMS. The predefined macro is called __VAMS_ENABLE__.

Example:

```
`ifdef __VAMS_ENABLE__
    parameter integer del = 1 from [1:100];
`else
    parameter del = 1;
`endif
```

When the *`resetall* compiler directive is encountered during compilation, all compiler directives are set to their default values. This is useful for ensuring that only those directives that are desired when compiling a particular source file are active. To do so, place *`resetall* at the beginning of each source text file, followed immediately by the directives desired in the file.

The *`timescale* compiler directive defines the time unit and the time precision for the modules that follow it. The time unit and time precision is specified using either 1, 10, or 100 followed by a measurement unit of either s, ms, us, ns, ps, or fs, which represents seconds, milliseconds, microseconds, nanoseconds, picoseconds, or femptoseconds.

Example:

```
`timescale 10ns / 1ns
```

The first value given specifies the units of time and the second specifies the precision. The values affect the way delays are specified and the return value from the $realtime function. Both are rounded to the time resolution and given in multiples of the time unit. Thus, with the specification given in the example above, #55.79 corresponds to a delay of 558ns (55.79 × 10 ns rounded to the nearest 1 ns).

2 Data Types

This section starts with a discussion of the various types of constants and variables available in Verilog-A/MS, and then presents signal types, including a discussion of natures and disciplines.

2.1 Constants

2.1.1 Integers

Underscores are ignored in numbers, so 42_839 is equivalent to 42839.

Examples: 124, +124, –124, 42_839

Except in Verilog-A, integer constants can be expressed in decimal, hexadecimal, octal, or binary. To do so, use *sb'fn*; where *s* is an optional sign, either '+' or '–'; *b* is an optional decimal number that indicates the size of the constant in bits; *f* is the base format and is either 'd', 'h', 'o', or 'b' for decimal, hexadecimal, octal, or binary; and *n* is the number in the specified base. In hexadecimal numbers the letters 'a' through 'f' represent the digits 10 through 15. Letters in integer constants can be either lower or upper case.

Examples:

63	*unsized decimal number*
'd63	*unsized decimal number*
'h3f	*unsized hexadecimal number*
'o77	*unsized octal number*
'b11_1111	*unsized binary number*
12'h3f	*12 bit hexadecimal number*
–'h3f	*negative unsized hexadecimal number*

The letters 'x' and 'z' can be given to denote unknown and high impedance digits in all but decimal numbers, and '_' is ignored. Sized constants for which the size is larger than the given number are padded on the left with zeros unless the first digit of the given number is an *x* or *z*, which are padded with the *x* or *z*. The number is truncated on the left if the size is smaller than the given number.

Examples:

12'hx	*a 12 bit unknown hexadecimal number*
64'o0	*a 64 bit octal 0 (zero padded)*
8'hfx	*equivalent to 8'b1111_xxxx*
8'hfffx	*equivalent to 8'b1111_xxxx (truncated)*
8'hx	*equivalent to 8'bxxxx_xxxx (x padded)*

2.1.2 Reals

Real numbers must either include a decimal point or a scale factor. If a decimal point is present, there must be digits on both sides. So .12, 9., 4.eE3, and .2e–7 are not valid numbers. Underscores are ignored in real numbers. Scale factors are given in Table 2.

Examples: 3.14, 0.1, 1.2E12, 1.30e–2, 236.123_763_e–12, 1.3u, 5.46K

Predefined numbers in the form of compiler directives are included in the file *constants.vams* and listed in Table 3. Mathematical constants are denoted with a `M_ prefix and physical constants use the `P_ prefix.

TABLE 2 *Scale factors for real numbers.*

Multiplier	Name	Symbol	Multiplier	Name	Symbol
10^{12}	tera	T	10^{-3}	milli	m
10^{9}	giga	G	10^{-6}	micro	u
10^{6}	mega	M	10^{-9}	nano	n
10^{3}	kilo	K or k	10^{-12}	pico	p
			10^{-15}	fempto	f
10^{d}	exponent	e*d* or E*d*	10^{-18}	atto	a

2.1.3 Strings

Strings are a sequence of characters enclosed in double quotes. Table 4 lists the escape sequences used to enter special characters into strings.

Example: "Hello World!\n"

2.1.4 Vectors

A ***constant vector*** is created using the ***concatenate*** operator, which consists of balanced braces surrounding a sequence of arguments given as expressions. It simply combines its arguments into an array. The individual arguments may be scalars or vectors, and the end result is a vector whose length equals the sum of the lengths of each argument.

Examples:

{4, 8, 12, 16, 20}
{4, 2*4, 3*4, 4*4, 5*4}
{4.0, 8.0, {12.0, 16.0, 20.0}}

In addition, the ***replicate*** operator can be used to specify a sequence of repeated values. The replication operator is similar to the concatenation operator, except the leading brace is preceded with an integer count and then the whole construct is surrounded with another set of braces. So {0, {2{1, 2}}} is equivalent to {0, 1, 2, 1, 2}

Vectors come in many forms. The examples above are numeric vectors, which can consist of either integers or real numbers. One can also have vectors of bits, nets, branches, instances and registers (referred to as memories).

TABLE 3 *Predefined constants in constants.vams*

Mathematical Constants		
`M_PI	π	3.14159265358979323846
`M_TWO_PI	2π	6.28318530717958647652
`M_PI_2	$\pi/2$	1.57079632679489661923
`M_PI_4	$\pi/4$	0.78539816339744830962
`M_1_PI	$1/\pi$	0.31830988618379067154
`M_2_PI	$2/\pi$	0.63661977236758134308
`M_2_SQRTPI	$2/\sqrt{\pi}$	1.12837916709551257390
`M_E	e	2.7182818284590452354
`M_LOG2E	$\log_2 e$	1.4426950408889634074
`M_LOG10E	$\log_{10} e$	0.43429448190325182765
`M_LN2	$\log_e 2$	0.69314718055994530942
`M_LN10	$\log_e 10$	2.30258509299404568402
`M_SQRT2	$\sqrt{2}$	1.41421356237309504880
`M_SQRT1_2	$1/\sqrt{2}$	0.70710678118654752440
Physical Constants		
`P_Q	charge of an electron	$1.602176462\times10^{-19}$ C
`P_C	speed of light	2.99792458×10^{8} m/s
`P_K	Boltzmann's constant	1.3806503×10^{-23} J/K
`P_H	Planck's constant	6.626076×10^{-34} J-s
`P_EPS0	permittivity of a vacuum	$8.854187817\times10^{-12}$ F/m
`P_U0	permeability of a vacuum	$\pi\times 4.0\times10^{-7}$ H/m
`P_CELSIUS0	0 Celsius	273.15 K

2.2 Variables

Variables can be thought of as named registers that contain a value of a particular type. They are initialized at the beginning of simulation to either zero or unknown as appropriate and cannot be explicitly initialized when declared. They retain their value until changed by way of an assignment statement. As such, they are different from variables in programming languages such as C in that they retain their value even when the flow of execution appears to leave their context (5§6.2.1p197).

TABLE 4 *Escape sequences for strings.*

Escape	Result
\n	new line
\t	tab
\\	\
\"	"
\ddd	Character represented by the 3 digit octal code *ddd* where ($0 \le d \le 7$)

A register or ***reg*** declaration declares arbitrarily sized logic variables (registers are not supported in Verilog-A). The default size is one bit.

Examples:

```
reg enable;
reg [15:0] bus;
```

In these examples, *enable* is a one bit variable and *bus* is a 16 bit variable. The index of the most significant bit is given first in the range specification, and the index of the least significant bit is given last. Any valid decimal integer may be given as an index bound.

Logic variables hold logic values. A one bit logic variable can be one of 4 possible values, shown in Table 5. Logic variable (registers) are initialized to *x*.

TABLE 5 *Verilog logic values.*

Value	Description
0	Zero, low, or false.
1	One, high, or true.
x or *X*	Unknown or uninitialized.
z or *Z*	High impedance (floating).

An ***integer*** declaration declares one or more variables of type integer. These variables can hold values ranging from -2^{31} to $2^{31}-1$. Arithmetic operations performed on integer variables produce 2's complement results. Integers are initialized at the start of a simulation depending on how they are used. Integer variables whose values are assigned in an analog process default to an initial value of zero (0). Such variables are said to be captured by the analog kernel. Integers that are captured by the analog kernel can only hold valid numbers (they may not contain any *x*- or *z*-valued bits). Integer variables whose values are assigned in a digital context default to an initial value of *x*. These variables are said to be captured by the discrete kernel. Such integers are implemented as 32-bit regs. As such, the values they hold may contain bits that are *x* or *z*.

Example: **integer** count, ub;

A ***real*** declaration declares one or more variables of type real. The real variables are stored as 64-bit quantities, as described by IEEE STD-754-1985, an IEEE standard for double precision floating point numbers. Real variables are initialized to zero (0) at the start of a simulation.

Example: **real** save, midpoint;

Arrays of integers and reals can be declared using a range that defines the upper and lower indices of the array. Both indices are specified with constant expressions that may evaluate to a positive integer, a negative integer, or to zero.

Example: **real** result[0:7];

A type of integer, ***genvar***, has restricted semantics that allow it to be used in static expressions. A *genvar* can only be assigned within the control section of a for loop. Assignments to the genvar variable can consist only of expressions of static values (expression involving only parameters, literal constants, and other *genvar* variables).

Example: **genvar** i;

2.2.1 Vectors

Individual members of a vector can be accessed by applying an index to a vector. An index is applied by following the identifier of the vector with an expression enclosed in balanced brackets. For example, x[3] accesses member 3 of the vector *x* (whether this is actually the third member of the vector depends on how *x* is declared).

Indexing of bit vectors, either in the form of vector registers or vectors of the traditional Verilog net types (see Table 8 on page 165) is often referred to as a ***bit-select*** process. In addition, bit vectors have an additional functionality not associated with other types of vectors that is referred to as a ***part-select*** process. In this case, a range of indices can be specified by placing a pair of expressions within the brackets separated by a colon. For example, x[0:3] accesses members 0 through 3 of *x*.

2.3 Parameters

Parameters are declared with a statement of the form

parameter <**real|integer**> *name=expr* <*rangeLimit*>;

The *parameter* keyword is followed by an optional type, either *real* or *integer.* The name of the parameter is followed by an initializing expression that when evaluated gives the default value of the parameter. More than one parameter of the same type can be declared in the same statement by adding more names (with initializer and optional range limit). If the type is not given explicitly, it is inferred from the type of the initializing expression. The range limit takes the form of one or more of the following

<**from|exclude**> [*lbExpr*:*ubExpr*] $lbExpr \le parameter \le ubExpr$
<**from|exclude**> (*lbExpr*:*ubExpr*] $lbExpr < parameter \le ubExpr$
<**from|exclude**> [*lbExpr*:*ubExpr*) $lbExpr \le parameter < ubExpr$
<**from|exclude**> (*lbExpr*:*ubExpr*) $lbExpr < parameter < ubExpr$
<**exclude**> *expr* $parameter \neq expr$

When the expressions are given as a pair separated by a colon, the first expression (*lbExpr*) is the lower bound of the range, and the second (*ubExpr*) is the upper bound.

Examples:

```
parameter bits=8, vdd=3.0;
parameter thresh=vdd/2;
parameter integer size=16;
parameter real td=0;
parameter integer bits = 8 from [1:24];
parameter integer dir = 1 from [-1:1] exclude 0;
parameter real period=1 from (0:inf);
parameter real toff=0 from [0:inf), td=0 exclude 0;
parameter real Vmin=0;
parameter real Vmax=Vmin+1 from (Vmin:inf);
```

Expressions must be written in terms of either literal constants or previously defined parameters.

Parameter values on instances of modules can be overwritten either by specifying a value when instantiating the module (5§9.2p227), or by using a *defparam* statement (5§9.4.2p233).

Parameters can also be arrays, in which case the array bounds are given after the parameter name and the parameter is initialized using an array. If the array size is changed via a parameter assignment, the parameter array must be assigned an array of the new size from the same module as the parameter assignment that changed the parameter array size.

Example:

```
parameter real poles[0:3] = {1.0, 3.198, 4.554, 12.0};
```

2.4 Natures and Disciplines

A ***nature*** is a collection of attributes that are shared by a class of signals. The attributes include the units (*units*), name used when accessing the signal (*access*), absolute tolerance (*abstol*), related natures (*ddt_nature*, *idt_nature*), and perhaps user or implementation defined attributes. Table 6 lists the natures included in the file *disciplines.vams*.

TABLE 6 *Natures available from disciplines.vams.*

Name	Units	Access	Abstol	Abstol Override
Voltage	V	V	10^{-6}	VOLTAGE_ABSTOL
Current	A	I	10^{-12}	CURRENT_ABSTOL
Charge	coul	Q	10^{-14}	CHARGE_ABSTOL
Flux	Wb	Phi	10^{-9}	FLUX_ABSTOL
Magneto_ Motive_Force	A-turns	MMF	10^{-12}	MAGNETO_MOTIVE_FORCE_ ABSTOL
Temperature	K	Temp	10^{-4}	TEMPERATURE_ABSTOL
Position	m	Pos	10^{-6}	POSITION_ABSTOL
Velocity	m/s	Vel	10^{-6}	VELOCITY_ABSTOL
Acceleration	m/s^2	Acc	10^{-6}	ACCELERATION_ABSTOL
Impulse	m/s^3	Imp	10^{-6}	IMPULSE_ABSTOL
Force	N	F	10^{-6}	FORCE_ABSTOL
Angle	rads	Theta	10^{-6}	ANGLE_ABSTOL
Angular_ Velocity	rads/s	Omega	10^{-6}	ANGULAR_VELOCITY_ABSTOL
Angular_ Acceleration	rads/s^2	Alpha	10^{-6}	ANGULAR_ACCELERATION_ ABSTOL
Angular_ Force	N-m	Tau	10^{-6}	ANGULAR_FORCE_ABSTOL

Example:

```
nature Current
    units = "A";
    access = I;
    abstol = 1p;
    idt_nature = Charge;
endnature
```

The absolute tolerances on the predefined natures can be overwritten by using *`define* to create a macro with the appropriate name (also given in Table 6) with the desired value.

Example:

```
`define VOLTAGE_ABSTOL 1e-3
`define CURRENT_ABSTOL 1e-9
`include "disciplines.vams"
```

Natures can also be derived from other natures, either directly or through a discipline. When doing so it is also possible to override the attributes.

Examples:

```
nature HighVoltage : Voltage
    abstol = 1m;
endnature

nature HighCurrent : electrical.flow
    abstol = 1n;
endnature
```

A ***discipline*** is a type used when declaring analog nodes, ports, or branches. They can also be used to declare digital wires and registers. A discipline may include the specification of a domain, either *continuous* or *discrete*, and up to two natures. At least one nature is required for continuous disciplines, for the potential. Continuous disciplines with a single nature are referred to as signal-flow disciplines. Conservative disciplines would also have a nature for the flow. Table 7 lists the disciplines available from *disciplines.vams.*

Example:

```
discipline electrical
    domain continuous;
    potential Voltage;
    flow Current;
enddiscipline
```

Continuous time signals belong to the *continuous* domain, whereas digital and discrete-event signals belong to the *discrete* domain. Signals in the continuous domain

TABLE 7 *Disciplines available from disciplines.vams.*

Name	Potential	Flow	Domain
logic*	—	—	discrete
electrical	Voltage	Current	continuous
voltage	Voltage	—	continuous
current†	Current	—	continuous
magnetic	Magneto_Motive_Force	Flux	continuous
thermal	Temperature	Power	continuous
kinematic	Position	Force	continuous
kinematic_v	Velocity	Force	continuous
rotational	Angle	Angular_Force	continuous
rotational_velocity	Angular_Velocity	Angular_Force	continuous

* The logic discipline is not available in Verilog-A.

† This represents a flaw in the language as presently defined in the LRM [28]. The language was inappropriately restricted so that if there was only one nature in a discipline, it must be the potential. This forces the current in a signal-flow current discipline to be represented as a potential, which makes it incompatible with the electrical discipline.

are real valued; signals in the discrete domain can either be binary (0,1, x, or z), integer or real valued. Unless otherwise specified, the domain of a discipline is taken to be *continuous*.

The attributes of a nature can be overridden from within a discipline.

Example:

```
discipline cmos
    potential Voltage;
    potential.abstol = 10u;
    flow Current;
    flow.abstol = 100p;
enddiscipline
```

2.4.1 Compatible Disciplines

Natures associated with the same base nature are compatible. Disciplines are compatible when their corresponding natures are compatible. Nodes, ports, and branches can be connected together only if their disciplines are compatible, and the two terminals on each of a branch must have compatible disciplines.

At each node there may be many different values of the absolute tolerance *abstol.* This may be because the various ports connected to an undeclared node have different, yet compatible, natures for either the potential, the flow, or both. Even if the natures are identical, the value of *abstol* may be overridden in the discipline of one or more of the ports. In such cases, all of the absolute tolerances must be satisfied at the node. This is equivalent to simply applying the smallest tolerance value for all calculations involving such nodes.

2.4.2 Empty Disciplines

It is possible to define a discipline with no natures. These are known as *empty disciplines* and they can be used in structural descriptions to let the components connected to a net determine which natures are to be used for the net.

Such disciplines may have a domain binding or they may be domain-less, allowing the domain to be determined by the connectivity of the net (4§4.2p123).

Example:

```
discipline interconnect
    domain continuous;
enddiscipline
```

2.4.3 Discipline of Wires and Undeclared Nets

A module can have nets that are undeclared. It might also have discrete nets declared without disciplines. They might be undeclared if they are bound only to ports in module instantiations. They might be declared without disciplines if they are declared using wire types (*wire*, *tri*, *wand*, *wor*, etc.). In these cases, the net is treated as having an empty discipline. If the net is referenced in behavioral code, then it is treated as having an empty discipline with a domain binding of *discrete*, otherwise it is treated as having an empty discipline with no domain binding. In these cases the actual discipline used for the net is determined by discipline resolution based on what is connected to the net.

2.4.4 Discipline Resolution

A node can consist of several nets, each existing in different modules and perhaps each with their own discipline declarations (5§2.5p164). If within a module a net is undeclared, it would take its discipline from that to which it is connected. Occasionally with discrete nets it is not possible to determine a discipline for a net by what it is connected to or what accesses it, in which case it is assigned the default discipline (as specified by `default_discipline), or if no default is defined, the empty discipline.

If everything that connects to an undeclared net has the same discipline, then the net will take that discipline. If it connects to ports of different, yet compatible, disciplines, then some additional information is needed to resolve the discipline of the net. However, in the case where the net does not cross domains, it is not necessary to actually resolve the discipline. For discrete nets the discipline is only used when inserting connect modules (4§4.3p128). For undeclared continuous nets the discipline specifies the units and tolerances of the net. Since only ports with compatible continuous disciplines can be connected to the same net the units are not an issue as compatible disciplines all have the same units. The tolerances are resolved, as stated in Section 2.4.1, by simply applying all of the tolerances to the net.

There are several situations where it is either desirable or necessary to fully resolve the discipline of an undeclared net. For the cases where all the disciplines associated with a net are compatible, it might be desirable if the default behavior of using the tightest tolerance from any discipline associated with the net is not appropriate. However, in cases where the disciplines are incompatible, the node must be split and connect modules inserted to link the now distinct parts of the newly partitioned node (4§4.3p128). In this case, the discipline must be fully resolved for each of the new nodes as the resolved discipline is needed when determining the type of connect module that is inserted. A continuous time node cannot be split in this manner, and so all nets of a continuous-time node must be compatible.

The discipline of an undeclared net can be fully resolved with any one of the following methods:

1. One could explicitly declare the discipline of the net locally.
2. One could explicitly declare the discipline of the net remotely using hierarchical names (5§9.4p230). For example

   ```
   electrical regulator.drive
   ```

   would specify the net *drive* in the instance *regulator* should have a discipline of *electrical*.
3. One could provide connect rules that direct the simulator on how to resolve disciplines when confronted with particular sets of disciplines on a net (4§4.2p123).

The process of resolving compatible disciplines with the help of ***connect rules*** (not available in Verilog-A) is illustrated in Figure 1. The process starts at the leaf level modules and proceeds up through the hierarchy to the root or the top-level module. In this example it is assumed that all disciplines are compatible. The first resolution step occurs in *Instance1* where the discipline of *net1* must be resolved. This net is assumed to be undeclared in *Instance1*, but it is connected to a port of *Instance11*, which has a discipline of *discipline11*, and to a port of *Instance12*, which has a discipline of *discipline12*. The first connect rule is used to resolve this net to a discipline of *discipline1*. This same procedure is used in the top-level module to resolve *net0*, and the node itself, to a discipline of *discipline0*. This process is described more fully for the case of mixed nets in Chapter 4 in Section 4.2 on page 123.

FIGURE 1 *Discipline resolution for compatible disciplines.*

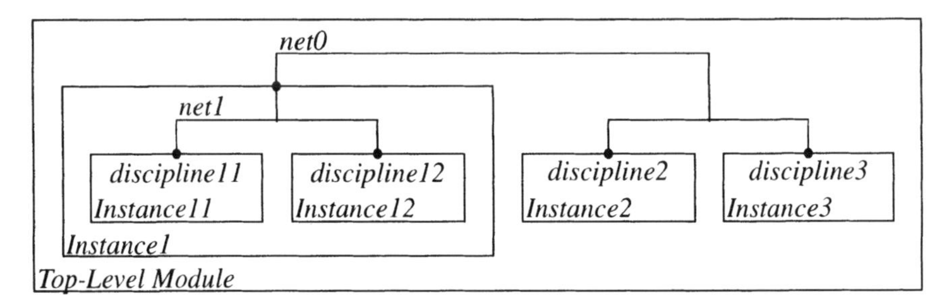


```
connectrules example;
    connect discipline11, discipline12 resolveto discipline1;
    connect discipline1, discipline2, discipline3 resolveto discipline0;
endconnectrules
```

2.5 Ports, Nets, and Nodes

A ***node*** is an electrically infinitesimal point of interconnection. A ***net*** is the name used for a node within a particular module. A ***port*** is a net that traverses the boundary of a module to the next higher level of the hierarchy. In Verilog-A nets and ports are declared using disciplines. With Verilog-AMS they may also be declared using the Verilog wire types, described in Table 8, or the new ***wreal*** type, which are discrete domain real-valued nets. Either scalar or vector nets may be declared.

Examples:

```
voltage p, n;
electrical [12:1] out;
logic [0:15] bus;
```

Ports are nets that are listed in a port list for the module. Each port must also have its directionality specified as *input*, *output* or *inout* (bidirectional). In Verilog-A a port must be explicitly declared net with a discipline. In Verilog-AMS, it may also be a net with either an explicitly or implicitly declared wire, or a register if the directionality is

TABLE 8 *Traditional Verilog net types (not available in Verilog-A).*

Type	Description
wire, tri	Value of wire resolves to the value of the strongest driver. If the strongest drivers have conflicting values, the value of the wire resolves to *x*. Drivers with value *z* have no strength.
wand, triand	Models a wired-and. The wire resolves to 0 if the value of any driver is 0.
wor, trior	Models a wired-or. The wire resolves to 1 if the value of any driver is 1.
supply0	Always resolves to a value of 0.
supply1	Always resolves to a value of 1.
tri0	When not driven or driven with *z* value, *tri0* has a value of 0.
tri1	When not driven or driven with *z* value, *tri1* has a value of 1.
trireg	Behaves like a wire except when the value of all drivers is *z*, in which case it retains its last driven value.

output. Other types of variables, reals and integers in particular, cannot be declared as ports, regardless of the directionality.

Example:

```
module amplifier (out, in);
    input in;
    output out;
    electrical out, in;
```

As of the most recent Verilog-AMS standard it is possible to combine the direction and type declaration into a single statement [28]. However, this feature is not yet available in most implementations.

Examples:

```
    input voltage in;
    output voltage out;
```

A vector net (a bus) is declared by providing a range immediately after the discipline or wire type (for this reason, nets with different sizes must be declared using separate statements). The range consists of the integer indices for the first and last members of the array.

An *input* port cannot be declared with type *reg* and *inout* ports cannot be *reg* or *wreal* or have a signal-flow discipline. The value of an input port cannot be assigned a value or be the target of a contribution statement within the module. The declaration of the direction of vector ports is given in a manner similar to the way their discipline is declared, though the dimension is optional in the direction statement.

Examples:

```
output [12:1] out;
electrical [12:1] out;

output out;
electrical [12:1] out;
```

The type of nets and ports need not be declared. Undeclared nets and ports used in the structural part of a module take on the type of the ports that connect to them (5§2.4p159). If they are used in the behavioral part of a module, they are assumed to be scalar wires. As the traditional Verilog wire types are not supported in Verilog-A, all nets and ports used in a Verilog-A behavioral description must be declared.

```
src #(.ampl(1)) src1 (n);
load #(.r(50) load1 (n);
```

In this example, *n* is an implicitly declared net.

The reference or ground node can be accessed by using the ***ground*** statement to give it a name that can be used within a module.

Example:

```
ground gnd;
```

When declaring nets and ports using the traditional Verilog net types, there are additional features available. One can add a delay to the net, which effectively adds delay to any driver of the net. In the following example, a delay of 10 time units (as specified with *`timescale* (5§1.4p151)) is added to the vector wire.

Example:

```
wire [7:0] #10 data;
```

One can also specify that individual members of a vector wire are inaccessible, that the members can only be accessed as a group. To do so, add the *vectored* keyword to the declaration. Conversely, to explicitly declare that members of a vector wire can be accessed individually using the bit-select and part-select mechanisms, add the *scalared* keyword (this is the default behavior).

Example:

```
wire vectored [7:0] #10 data;
wire scalared [5:0] #10 control;
```

Finally, it is possible to declare a logical expression that always acts to drive a net when declaring the net. This is referred to as net assignment (5§7.4.3p214).

Example:

```
wire c = a ~& b;
```

2.6 Branches

A ***branch*** is a path between two nets. Each branch is associated with the two nets from which it derives a discipline. These nets are referred to as the ***branch terminals***. Only one net need be specified, in which case the second net defaults to ground and the discipline for the branch is derived from the specified net. The disciplines for the specified nets must be compatible (5§2.4.1p161).

Branches are explicitly declared and named using

```
branch (n1, n2) branch1, branch2;
```

In this case, two branches are created, both of which are connected between nets *n1* and *n2*. If either of the nets is a vector net, then the branch becomes a vector branch. If both nets are vectors, they must have the same size.

In addition to explicitly named and declared branches, implicitly declared unnamed branches are also supported. In this case, a pair of nets is used to identify the branch. (5§3.1p168)

3 Signals

Signals are values that vary with time that are passed between modules. There are two types of signals, continuous time (often referred to as analog) and discrete event (often referred to as digital, but also includes continuous-value (analog) discrete-event signals). Discrete event signals belong to the discrete domain, meaning that they are associated (owned) by the discrete-event kernel. They are piecewise constant or 'event driven' and may or may not be associated with explicitly declared disciplines. All continuous-time signals belong to the continuous domain and are associated with explicitly declared disciplines (5§2.4p159). Continuous-time signals are associated with the continuous-time kernel and their value can only be accessed through the use of access functions, which implies that they must have disciplines.

3.1 Continuous-Time Signal Access

3.1.1 Access Functions

Flow and potential signals on nets, ports, and branches are accessed using ***access functions***. The name of the access function is taken from the discipline of the net, port, or branch associated with the signal. The list of predefined access functions was given earlier in Table 6 on page 159. For the *electrical* discipline, the access functions are *V* and *I*, with *V* accessing the voltage (the potential) and *I* accessing the current (the flow).

In the following examples, assume that *n* and *m* are either nets or ports, *b* is a branch, and *p* is a port, and that all are associated with the *electrical* discipline.

Examples:

```
V(n)        // Voltage from n to ground (unnamed branch)
I(n)        // Current from n to ground (unnamed branch)
V(b)        // Voltage across b (named branch)
I(b)        // Current through b (named branch)
V(n,m)      // Voltage from n to m (unnamed branch)
I(n,m)      // Current from n to m (unnamed branch)
I(<p>)      // Current through p (port branch)
```

In all cases, one can say that the access function is applied to a branch, where the branch can be an unnamed branch, a named branch, or a port branch.

An access function can only be applied to a scalar branch. Thus, it can be applied to the member of a bus, but not the bus itself. When applying an access function to a member of a bus, the index must be a *genvar* expression (5§2.2p155).

Example:

```
V(a[2])
```

To read the value of a signal, apply an access function to a branch to gain access to the desired signal, and place that access function in an expression. To modify the value of a signal, the access function must be the target of a contribution operator.

Example:

```
I(cap) <+ c*ddt(V(cap));
```

3.1.2 Accessing Signal Attributes

Attributes are attached to the nature of a potential or flow. The attributes for a net or a branch can be accessed by using the hierarchical referencing operator (.) to the *potential* or *flow* for the net or branch. For example, if *a* is a net, port, or branch, then *a.potential.abstol* is the *abstol* for the potential of *a*.

Example:

```
analog begin
    in = max(V(p), V(n));
    integ = idt(in, p.potential.abstol);
    V(out) <+ abs(integ);
end
```

3.2 Contributions

Analog signals (potentials and flows) are generally assigned values within contribution statements (also described in (5§6.4p198) and in Chapter 3 in Section 4 starting on page 58). Contribution statements always contain a contribution operator, denoted '<+'. Before the contribution operator is an access function, defining the target of the contribution. After the contribution operator is an expression, which evaluates to the value of the contribution.

Example:

```
V(out) <+ transition( quantized, td, tt );
```

If there are multiple contributions to the same branch within the same analog process, the contributions accumulate.

Example:

```
I(diode) <+ is*(limexp(V(diode)/$vt) – 1);
I(diode) <+ ddt(–2*cjo*phi*sqrt(1 – V(diode)/phi));
```

In this example the evaluated value of both expressions is added to the branch.

An important feature of contribution statements is that the value of the target may be expressed in terms of itself. This is referred to as an implicit or fixed-point formulation.

Example:

```
I(diode) <+ is*(limexp(V(diode)/$vt – r*I(diode)) – 1);
```

Notice that I(diode) is found on both sides of the contribution operator. The underlying implementation of the simulator will find the value of I(diode) that equals the sum of the contributions made to it, even if the contributions are a function of I(diode) itself. This feature is demonstrated in Listing 12 on page 59 of Chapter 3.

3.2.1 Switch Branches

At any point in time one cannot contribute to both the potential and the flow of a branch. However, it is possible to change the target of the contribution over time from potential to flow and back again. In doing so, one is said to be contributing to a ***switch branch***.

Example:

```
if (V(ps,ns) > thresh)
    V(p,n) <+ 0;
else
    I(p,n) <+ 0;
```

In addition, a switch branch is also created if the potential of the branch is the target of a contribution statement, but the statement is not evaluated at every point in time. In this case, when no contribution is made, the branch flow is zero.

Example:

```
if (V(ps,ns) > thresh)
    V(p,n) <+ 0;
```

This example is equivalent to the previous example.

3.2.2 Indirect Branch Assignments

Contribution statements are not the only way that values can be assigned to analog signals. Indirect branch assignments provide an alternative approach that is useful in cases where contributions do not behave as needed. Once such case is the ideal opamp (or nullor). In this model, the output is driven to the voltage that results in the input voltage being zero. The constitutive equation is

$$v_{in} = 0, \tag{1}$$

which can be formulated with a contribution statement as

```
V(out) <+ V(out) + V(in);
```

This statement defines the output of the *opamp* to be a controlled voltage source by assigning to V(out) and defines the input to be high impedance by only probing the input voltage. That the desired behavior is achieved can be seen by subtracting V(out) from both sides of the contribution operator, resulting in (1). However, this approach does not result in the right tolerances being applied to the equation if out and in have different disciplines.

The ***indirect branch assignment*** should be used in this situation.

```
V(out): V(in) == 0;
```

which reads 'drive V(out) so that V(in) == 0'. This indicates out is driven with a voltage source and the source voltage needs to be adjusted so that the given equation is satisfied. Any branches referenced in the equation are only probed and not driven. In particular, V(in) acts as a voltage probe.

The left hand side of the equality operator must either be an access function, or ddt or idt applied to an access function. The tolerance for the equation is taken from the argument on the left side of the equality operator.

An application of the indirect branch assignment is shown in Listing 1, a module that describes an ideal opamp.

LISTING 1 *An ideal opamp.*

```
module ideal_opamp (pout, nout, pin, nin);
    output pout, nout;
    input pin, nin;
    electrical pin, nin, pout, nout;
    branch (pout,nout) out;
    branch (pin,nin) in;

    analog begin
        V(out): V(in) == 0;
    end
endmodule
```

3.2.3 Multiple Indirect Assignments

For multiple indirect assignments statements, the targets frequently can be paired with any equation.

Example:

The following system of ordinary differential equations,

$$\frac{dx}{dt} = f(x, y, z), \tag{2}$$

$$\frac{dy}{dt} = g(x, y, z), \tag{3}$$

$$\frac{dz}{dt} = h(x, y, z), \tag{4}$$

can be written as

```
V(x): ddt(V(x)) == f(V(x), V(y), V(z));
V(y): ddt(V(y)) == g(V(x), V(y), V(z));
V(z): ddt(V(z)) == h(V(x), V(y), V(z));
```

or

```
V(y): ddt(V(x)) == f(V(x), V(y), V(z));
V(z): ddt(V(y)) == g(V(x), V(y), V(z));
V(x): ddt(V(z)) == h(V(x), V(y), V(z));
```

or

```
V(z): ddt(V(x)) == f(V(x), V(y), V(z));
V(x): ddt(V(y)) == g(V(x), V(y), V(z));
V(y): ddt(V(z)) == h(V(x), V(y), V(z));
```

without affecting the results.

3.2.4 Indirect Assignment and Contribution

Indirect assignment is incompatible with contribution. Once a value is indirectly assigned to a branch, it cannot be contributed to using the branch contribution operator. It is illegal to indirectly assign to an external branch or contribute to an external branch that has an indirect branch assignment.

4 Expressions

An ***expression*** is a construct that combines ***operands*** with ***operators*** to produce a result that depends on the values of the operands and the semantic meaning of the operators. Any legal operand without an operator is also considered an expression. Wherever a value is needed in a Verilog-AMS statement, an expression can be used.

Some statement constructs require an expression to be a ***constant expression***. The operands of a constant expression consist of literal numbers and parameter names, but they can use any of the operators defined in Table 9 or the functions in Table 10.

4.1 Operators

The operators available in Verilog-AMS are listed in Table 9.

4.2 Functions

A function takes a collection of arguments and returns a value based on the values of the arguments. The arguments are passed by order in a comma separated list from the expression where the function is called. In general, values are not required for all parameters as some parameters have default values. To indicate that the default value should be used, simply do not specify the argument value. If the argument value is expected at the end of the argument list, both the argument and the comma that sepa-

TABLE 9 *Verilog-AMS operators*

<table>
<tr><th>Symbol</th><th>Usage</th><th>Description</th><th>*</th><th>†</th></tr>
<tr><td colspan="5">Arithmetic Operators</td></tr>
<tr><td>+</td><td>a + b</td><td>Sum a and b</td><td>y</td><td>y</td></tr>
<tr><td>–</td><td>a – b</td><td>Subtract b from a</td><td>y</td><td>y</td></tr>
<tr><td>–</td><td>–a</td><td>Negate a</td><td>y</td><td>y</td></tr>
<tr><td>*</td><td>a * b</td><td>Multiply a and b</td><td>y</td><td>y</td></tr>
<tr><td>/</td><td>a / b</td><td>Divide a by b (b must be nonzero)</td><td>y</td><td>y</td></tr>
<tr><td>%</td><td>a % b</td><td>Modulus of a / b (b must be nonzero)</td><td>y</td><td>y</td></tr>
<tr><td colspan="5">Bitwise Operators</td></tr>
<tr><td>~</td><td>~ a</td><td>Invert each bit of a</td><td>n</td><td>y</td></tr>
<tr><td>&</td><td>a & b</td><td>Logical ‘and’ of each bit of a and b</td><td>n</td><td>y</td></tr>
<tr><td>~&</td><td>a ~& b</td><td>Logical ‘nand’ of each bit of a and b</td><td>n</td><td>y</td></tr>
<tr><td>|</td><td>a | b</td><td>Logical ‘or’ of each bit of a and b</td><td>n</td><td>y</td></tr>
<tr><td>~|</td><td>a ~| b</td><td>Logical ‘nor’ of each bit of a and b</td><td>n</td><td>y</td></tr>
<tr><td>^</td><td>a ^ b</td><td>Logical ‘xor’ of each bit of a and b</td><td>n</td><td>y</td></tr>
<tr><td>~^</td><td>a ~^ b</td><td>Logical ‘xnor’ of each bit of a and b</td><td>n</td><td>y</td></tr>
<tr><td>^~</td><td>a ^~ b</td><td>Same as ‘~^’</td><td>n</td><td>y</td></tr>
<tr><td colspan="5">Unary Reduction Operators</td></tr>
<tr><td>&</td><td>&a</td><td>Logical ‘and’ all bits in a to form 1 bit result</td><td>n</td><td>y</td></tr>
<tr><td>~&</td><td>~&a</td><td>Logical ‘nand’ all bits in a to form 1 bit result</td><td>n</td><td>y</td></tr>
<tr><td>|</td><td>|a</td><td>Logical ‘or’ all bits in a to form 1 bit result</td><td>n</td><td>y</td></tr>
<tr><td>~|</td><td>~|a</td><td>Logical ‘no’ all bits in a to form 1 bit result</td><td>n</td><td>y</td></tr>
<tr><td>^</td><td>^a</td><td>Logical ‘xor’ all bits in a to form 1 bit result</td><td>n</td><td>y</td></tr>
<tr><td>~^</td><td>~^a</td><td>Logical ‘xnor’ all bits in a to form 1 bit result</td><td>n</td><td>y</td></tr>
<tr><td>^~</td><td>^~a</td><td>Same as ‘~^’.</td><td>n</td><td>y</td></tr>
</table>

TABLE 9 *Verilog-AMS operators*

Symbol	Usage	Description	*	†
Logical Operators				
!	!*a*	Is *a* false (1 bit result)?	y	y
&&	*a* && *b*	Are both *a* and *b* true (1 bit result)?	y	y
||	*a* || *b*	Are either *a* or *b* true (1 bit result)?	y	y
Equality Operators‡				
==	*a* == *b*	Is *a* equal to *b* (1 bit result)?	y	y
!=	*a* != *b*	Is *a* not equal to *b* (1 bit result)?	y	y
Identity Operators				
===	*a* === *b*	Is *a* identical to *b* (1 bit result)?	n	n
!==	*a* !== *b*	Is *a* not identical to *b* (1 bit result)?	n	n
Relational Operators				
<	*a* < *b*	Is *a* less than *b* (1 bit result)?	y	y
>	*a* > *b*	Is *a* greater than *b* (1 bit result)?	y	y
<=	*a* <= *b*	Is *a* less than or equal to *b* (1 bit result)?	y	y
>=	*a* >= *b*	Is *a* greater than or equal to *b* (1 bit result)?	y	y
Logical Shift Operators				
<<	*a* << *b*	Shift *a* left *b* times, vacated bits are filled with 0	n	y
>>	*a* >> *b*	Shift *a* right *b* times, vacated bits are filled with 0	n	y
Miscellaneous Operators				
? :	*a* ? *b* : *c*	Evaluates to *b* if *a* is true, and *c* otherwise	y	y
{ }	{*a*, *b*}	Concatenates sized bit vectors *a* and *b* creating a larger bit vector	y	y
{ { } }	{a{b}}	Replicate *b*, *a* times.	y	y
->	-> *a*	Trigger a named event *a*	n	n

* Supports real operands.

† Available in Verilog-A.

‡ Operand values of *x* or *z* cause operator to evaluate to *x*.

rates it from the previous value are not specified. Otherwise the comma is given, but not the value.

4.3 Mathematical Functions

The standard mathematical functions supported by Verilog-A/MS are shown in Table 10. The arguments must be numeric (*integer* or *real*). For *min*(), *max*(), and *abs*(), if either argument is *real*, both are converted to *real*, as is the result. All arguments to the other functions are always converted to *real*. All arguments to the trigonometric and hyperbolic functions are specified in radians.

The *min*(), *max*(), and *abs*() functions have points where their derivatives are undefined. In order to define the behavior of the derivative of these functions at their points of discontinuity, these functions are defined as:

min(x,y) is equivalent to $(x < y)\ ?\ x : y$
max(x,y) is equivalent to $(x > y)\ ?\ x : y$
abs(x) is equivalent to $(x > 0)\ ?\ x : -x$

4.4 Logical Functions

A logical function is a function that returns either a 1 or 0 in the form of an integer to signify true or false, yes or no.

4.4.1 Analysis

The *analysis* function takes one or more string arguments and returns 1 (true) if any of the arguments match the current analysis type or name.

analysis(*str1*, <*str2*>, ...)

The names given in Table 11 are the names used for traditional Spice-like analyses. In addition, individual simulators may define their own names. Any unsupported names are simply assumed not to match.

4.5 Environment Functions

The environment functions return information about the current circuit environment in the form of a real number.

4.5.1 Current Time ($abstime)

The $*abstime* (short for absolute time) function takes no arguments and returns a real value that equals the current time in seconds.

TABLE 10 *Mathematical functions.*

Function	Description	Domain
ln(x)	Natural logarithm	$x > 0$
log(x)	Decimal logarithm	$x > 0$
exp(x)	Exponential	$x < 80$
sqrt(x)	Square root	$x \geq 0$
min(x, y)	Minimum	All x, all y
max(x, y)	Maximum	All x, all y
abs(x)	Absolute value	All x
floor(x)	Floor	All x
ceil(x)	Ceiling	All x
pow(x, y)	Power (x^y)	all y if $x \geq 0$; integer y if $x < 0$
sin(x)	Sine	All x
cos(x)	Cosine	All x
tan(x)	Tangent	$x \neq n(\pi/2)$, n is odd
asin(x)	Arc-sine	$-1 \leq x \leq 1$
acos(x)	Arc-cosine	$-1 \leq x \leq 1$
atan(x)	Arc-tangent	All x
atan2(x,y)	Arc-tangent of x/y	All x, All y
hypot(x,y)	$\sqrt{x^2 + y^2}$	All x, All y
sinh(x)	Hyperbolic sine	$x < 80$
cosh(x)	Hyperbolic cosine	$x < 80$
tanh(x)	Hyperbolic tangent	All x
asinh(x)	Arc-hyperbolic sine	All x
acosh(x)	Arc-hyperbolic cosine	$x \geq 0$
atanh(x)	Arc-hyperbolic tangent	$-1 \leq x \leq 1$

4.5.2 Current Time ($realtime)

The *$realtime* function returns a real value equal to the current time in time units (as specified with *`timescale* (5§1.4p151)).

TABLE 11 *Names for the SPICE-like analyses.*

Name	Analysis
"ac"	The *.ac* analysis
"dc"	The *.op* or *.dc* analyses.
"noise"	The *.noise* analysis.
"tran"	The transient, or *.tran*, analysis.
"ic"	The initial condition analysis that precedes a transient analysis.
"static"	Any equilibrium point calculation, including a DC analysis as well as those that precede another analysis, such as the DC analysis that precedes an AC or noise analysis, or the IC analysis that proceeds a transient analysis.
"nodeset"	The phase during an equilibrium point analysis where the nodesets are forced.

4.5.3 Ambient Temperature ($temperature)

The *$temperature* function takes no arguments and returns the current ambient temperature in Kelvin.

4.5.4 Thermal Voltage ($vt)

The *$vt* function returns the thermal voltage,

$$V_T = \frac{kT}{q}, \tag{5}$$

where k is Boltzmann's constant (`P_K), $T$ is the temperature in Kelvin as specified as the argument to the function, and $q$ is the charge of an electron, (`P_Q). If no argument is specified, the temperature is taken to be the current temperature as returned by *$temperature*.

4.6 Analog Operators

Analog operators operate on an expression that varies with time and return a value. They are functions that operate on more than just the current value of their arguments and so maintain internal state, with their output being dependent on both the input and the internal state. Analog operators are also sometimes referred to as filters.

Analog operators are subject to several important restrictions because they maintain their internal state.

- Analog operators must not be used inside conditional (*if, case*) or looping (*for*) statements unless the conditional expression controlling the statement consists of terms that cannot change their value during the course of an analysis, that is, unless *genvar* expressions are used to control the statement (3§10p84).
- Analog operators are not allowed in the *repeat* and *while* looping statements.
- Analog operators are not allowed in the body of an event statement (3§9p80).
- Analog operators can only be used inside an *analog process*; they cannot be used inside an *initial* or *always process*, or inside user-defined functions.
- It is illegal to specify a null operand argument to an analog operator.

These restrictions prevent usage that could cause the internal state to become corrupted or out-of-date. Table 12 lists the notable restrictions on the various analog operators and functions.

TABLE 12 *Analog operators and functions with notable restrictions.*

Operator	Restrictions
<+	Must be found within an analog process. Not permitted in event clauses, unrestricted loops, or function definitions.
@	Not permitted in event clauses or function definitions.
idt, ddt, idtmod	Must be found within an analog process. Not permitted within an event clause, an unrestricted conditional or loop, or function definitions.
*laplace_*, zi_**	Must be found within an analog process. Not permitted within an event clause, an unrestricted conditional or loop, or function definitions.
transition, slew	Must be found within an analog process. Not permitted within an event clause, an unrestricted conditional or loop, or function definitions.
cross, above	Must be found in an event expression. Not permitted within an event clause, an unrestricted conditional or loop, or function definitions.
last_crossing	Must be found within an analog process. Not permitted within an event clause, an unrestricted conditional or loop, or function definitions.

TABLE 12 *Analog operators and functions with notable restrictions.*

Operator	Restrictions
limexp	Must be found within an analog process. Not permitted within an event clause, an unrestricted conditional or loop, or function definitions.

4.6.1 Time Derivative (ddt)

ddt(*operand*, <*abstol*|*nature*>)

Returns the derivative of *operand* with respect to time. Takes an optional argument from which the absolute tolerance is determined. That argument is either the tolerance itself, or it is a nature from which the tolerance is extracted.

The output of a *ddt* operator during a quiescent operating point analysis is 0. During a small signal frequency domain analysis, such as AC or noise, the transfer function of the *ddt* operator is $j2\pi f$ where $j = \sqrt{-1}$ and f is the frequency of the analysis.

4.6.2 Time Integral (idt)

idt(*operand*, <*ic*>, <*assert*>, <*abstol*|*nature*>)

Returns the integral of *operand* with respect to time. Takes an initial condition, *ic*, that is asserted at the beginning of the simulation, and whenever *assert* is nonzero. Takes an optional argument from which the absolute tolerance is determined. That argument is either the tolerance itself, or it is a nature from which the tolerance is extracted (3§13p94).

During a DC operating point analysis the apparent gain from its input, *operand*, to its output is infinite unless an initial condition is supplied and asserted. If no initial condition is supplied, the *idt* function must be part of a negative feedback loop that drives its input value to zero, otherwise the simulator will fail to converge. During a small signal frequency domain analysis, such as AC or noise, the transfer function of the *idt* function is $1/j2\pi f$ where $j = \sqrt{-1}$ and f is the frequency of the analysis.

4.6.3 Circular Time Integral (idtmod)

idtmod(*operand*, <*ic*>, <*modulus*>, <*offset*>, <*abstol*|*nature*>)

Returns the integral of *operand* with respect to time. Takes an optional initial condition, *ic*, that if given is asserted at the beginning of the simulation. If the *modulus* is given, the output wraps so that it always falls between *offset* and *offset+modulus* as shown in Figure 2. The default value for *offset* is 0. Takes an optional argument from which the absolute tolerance is determined. That argument is either the tolerance itself, or it is a nature from which the tolerance is extracted (3§13p94).

FIGURE 2 *Output of the idtmod operator when input argument is a constant α, the initial condition is y_0, the modulus is m, the offset is b, and k is an integer.*

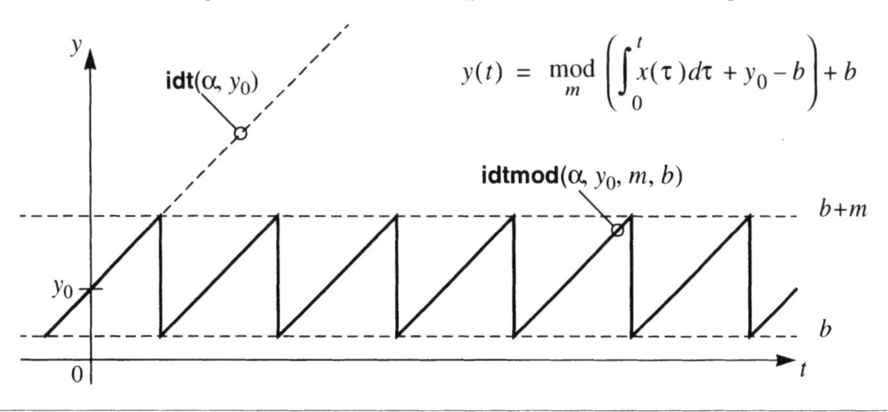

In DC analysis the *idtmod* function behaves the same as the *idt* function (except the *idt* output is passed through the modulus function). As such, the same warnings apply. The small signal frequency domain analysis behavior is the same as the *idt* function; the transfer function is $1/j2\pi f$.

4.6.4 Transition

transition(*operand*, <*delay*>, <*trise*>, <*tfall*>, <*ttol*>)

Converts a piecewise constant waveform, *operand*, into a waveform that has controlled transitions. The transitions have the specified *delay* and transition time (*trise* and *tfall*). If only *trise* is given, then *tfall* is taken to be the same as *trise*. If not specified, the transition times are taken to be the value of the currently active `default_transition compiler directive. The transition time is sometimes referred to as the inertial delay, in which case *delay* is referred to as the transport delay.

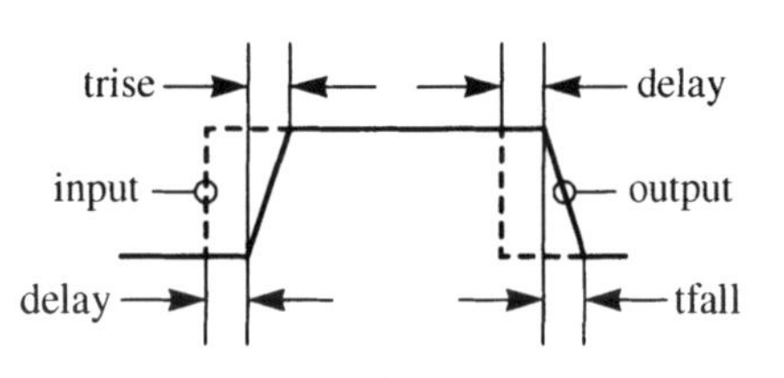

Normally the transition filter causes the simulator to place time points on each of the corners of the transition. However, if the transition time is specified to be zero, the transition occurs in the default transition time and no attempt is made to resolve the trailing corner of the transition. The time tolerance *ttol*, when nonzero, allows the times of the transition corners to be adjusted for better efficiency within the given tolerance.

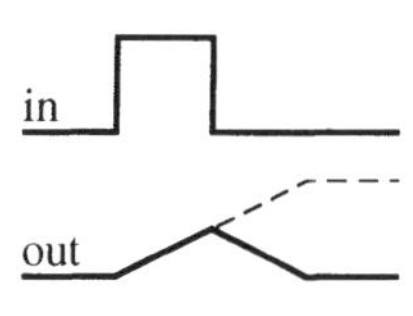

Since transitions take some time to complete, it is possible for a new output transition to be due to start before the previous transition is complete. In this case, the transition function terminates the previous transition and shifts to the new one in such a way that the continuity of the output waveform is maintained. Thus, the transition function naturally produces glitches or runt pulses. In addition, the transition filter internally maintains a queue of output transitions that have been scheduled but not processed. Each has an associated delay and transition time, which are the values of the associated arguments when the change in the value of the operand occurred. Since the delay can be different for each transition, it may be that the output from a change in the input may occur before the output from an earlier change. In this case, the transition that results from the change of the input that occurs later will preempt outputs from those that occurred earlier if their output occurs earlier.

During a DC operating point analysis the output of the *transition* function equals the value of *operand*. During a small signal analysis no signal passes through the *transition* function.

4.6.5 Slew

slew(*operand*, <*maxPosSlope*>, <*maxNegSlope*>)

Given an input waveform, *operand*, *slew* produces an output waveform that is the same as the input waveform except that it has bounded slope. The maximum positive slope and maximum negative slope are specified as arguments, *maxPosSlope* and *maxNegSlope*. If *maxNegSlope* is not specified, it is taken to be the same as *maxPosSlope*.

During a DC operating point analysis the output of the *slew* function will equal the value of *operand*. During a small signal analysis, such as AC or noise, the *slew* function will exhibit zero gain if slewing at the operating point and unity gain otherwise.

4.6.6 Delay

absdelay(*operand*, *delay*, <*maxDelay*>)

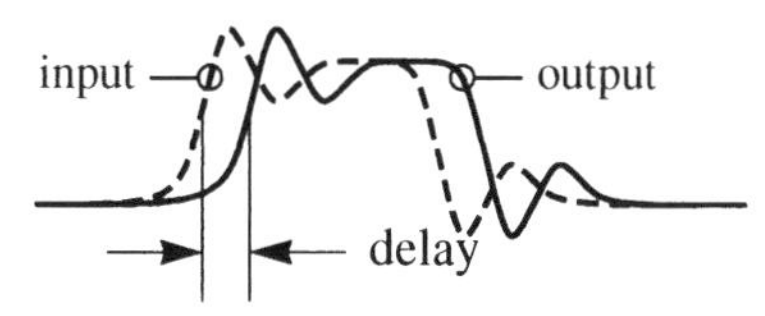

Returns a waveform that equals the input waveform, *operand*, delayed in time by an amount equal to *delay*, the value of which must be positive (the operator is causal). If *maxDelay* is specified, then *delay* is allowed to vary but must never be larger than *maxDelay*.

During a DC operating point analysis the output of the *absdelay* function will equal the value of *operand*. During a small signal frequency domain analysis, such as AC or noise, the transfer function of the *absdelay* function is $e^{j2\pi fT}$ where T is the value of the *delay* argument and f is the frequency of the analysis.

4.6.7 Laplace Transform Filters

The Laplace transform filters implement lumped linear continuous-time filters. Each filter takes a common set of parameters, the first is the input to the filter. The next two specify the filter characteristics. They are static, meaning they must not change during the course of the simulation. Finally, an optional parameter specifies the absolute tolerance. It may be a real number that directly gives the tolerance or a nature from which the tolerance is derived. Whether an absolute tolerance is needed depends on the context where the filter is used (3§13p94).

The Laplace transforms are written in terms of the variable s. The behavior of the filter in the time domain can be found by convolving the inverse of the Laplace transform with the input waveform. In frequency domain analyses, the transfer function is found by substituting $s = j2\pi f$, where $j = \sqrt{-1}$. For quiescent operating point analyses, such as a DC analysis, the transfer characteristics are found by setting $s = 0$ [25,30].

Laplace Transform Filter Functions. The filter *laplace_zp*() implements the zero-pole form of the Laplace transform filter. The general form is

laplace_zp(*operand*, <ζ>, ρ, <ε>)

where ζ (zeta) is a vector of M pairs of real numbers. Each pair represents a zero, the first number in the pair is the real part of the zero frequency (in radians per second) and the second is the imaginary part. The zeros argument is optional. Similarly, ρ (rho) is the vector of N real pairs, one for each pole. The poles are given in the same manner as the zeros. The transfer function is

$$H(s) = \frac{\prod_{k=0}^{M-1} 1 - \frac{s}{\zeta_k^R + j\zeta_k^I}}{\prod_{k=0}^{N-1} 1 - \frac{s}{\rho_k^R + j\rho_k^I}}, \tag{6}$$

where ζ_k^R and ζ_k^I are the real and imaginary parts of the kth zero, while ρ_k^R and ρ_k^I are the real and imaginary parts of the kth pole. If a root (a pole or zero) is real, the imaginary part is specified as zero. If a root is complex, its conjugate must also be present. If a root is zero, then the term associated with it is implemented as s, rather than $(1 - s/r)$ (where r is the root). If $\rho_k^R < 0$ then the kth pole is stable.

Example:

```
V(out) <+ laplace_zp(V(in), {-2, 0}, {-1, -3, -1, 3});
```

implements

$$\frac{V_{out}(s)}{V_{in}(s)} = \frac{1 + \frac{s}{2}}{\left(1 + \frac{s}{1 + 3j}\right)\left(1 + \frac{s}{1 - 3j}\right)}. \tag{7}$$

laplace_nd implements the rational polynomial form of the Laplace transform filter. The general form is,

laplace_nd(*operand*, *n*, *d*, <ε>);

where n is a vector of M real numbers containing the coefficients of the numerator and d is a vector of N real numbers containing the coefficients of the denominator. The transfer function is

$$H(s) = \frac{\sum_{k=0}^{M} n_k s^k}{\sum_{k=0}^{N} d_k s^k}. \tag{8}$$

Example:

```
V(out) <+ laplace_nd(V(in), {0, 3}, {4, 0, 1});
```

implements

$$\frac{V_{out}(s)}{V_{in}(s)} = \frac{3s}{s^2 + 4}. \tag{9}$$

The filters *laplace_zd* and *laplace_np* are similar to the Laplace filters already described with *laplace_zd* accepting a zero/denominator polynomial form and *laplace_np* taking a numerator polynomial/pole form.

laplace_zd(*operand*, ζ, *d*, <ε>)
laplace_np(*operand*, *n*, ρ, <ε>);

4.6.8 Z Transform Filters

The z transform filters implement lumped linear discrete-time filters. Since they exist within analog processes, their inputs and outputs are continuous-time waveforms. As shown in Figure 3, each filter function internally samples its input waveform $x(t)$ to form a sequence x_n, it filters that sequence to produce an output sequence y_n, and then it passes that sequence through a zero-order hold to produce $y(t)$. The input sampler is controlled by two parameters common to each of the filters, T and t_0. T is the sampling interval or time between samples and t_0 is the time of the first sample. The output zero-order hold is also controlled by two common parameters, T and τ. T is the total hold time for a sample and τ is the transition time, or the time the output takes to transition from one value to the next. During the transition, the output engages in a linear ramp between the old and new values so as to eliminate the discontinuous jump that would otherwise occur.

FIGURE 3 *The internal workings of a z-domain filter function.*

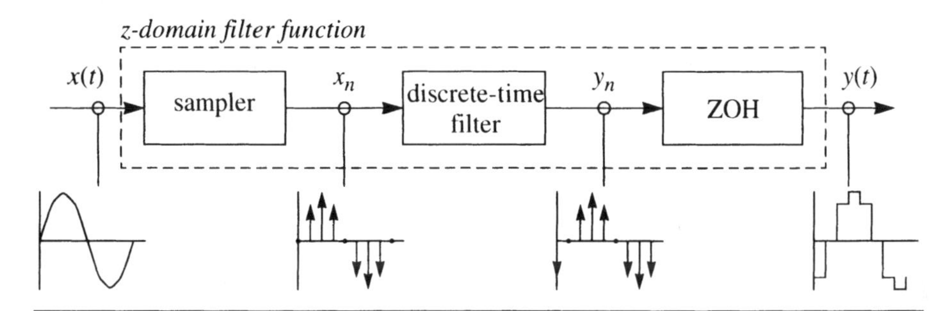

In addition to these three parameters, each z-domain filter takes three more arguments. The first is the input signal, $y(t)$. The other two are vectors that describe the z-domain transfer function of the discrete-time filter. As with the Laplace filters, the transfer function can be described using either the coefficients or the roots of the numerator and denominator polynomials. The filter characteristics are static, meaning that any changes that occur during the course of the simulation in the values contained within these vectors are ignored; only their initial values are important.

The z transforms are written in terms of the variable z. The behavior of the internal discrete-time filter in the time domain can be found by convolving the inverse of the z transform with the input sequence, x_n. The composite behavior then includes the effect of the sampler and the zero-order hold. In frequency domain analyses, the transfer function of the digital filter is found by substituting $z = e^{sT}$ where $s = j2\pi f$ and $j = \sqrt{-1}$. The composite frequency response would also include modifications to account for the effect of the sampler and zero-order hold. For quiescent operating point analyses, such as a DC analysis, the composite transfer characteristics are found by evaluating $H(z)$ for $z = 1$ [25,30].

Z Transform Filter Functions. The filter *zi_zp*() implements the zero-pole form of the z transform filter (*zi* is short for *z inverse*). The general form is

zi_zp(*operand*, <ζ>, ρ, *T*, <τ>, <t_0>)

where ζ (zeta) is a vector of M pairs of real numbers. Each pair represents a zero, the first number in the pair is the real part of the zero frequency (in radians per second) and the second is the imaginary part. The zeros argument is optional. Similarly, ρ (rho) is the vector of N real pairs, one for each pole. The poles are given in the same manner as the zeros. The transfer function is

$$H(z) = \frac{\prod_{k=0}^{M-1} 1 - \frac{z^{-1}}{\zeta_k^R + j\zeta_k^I}}{\prod_{k=0}^{N-1} 1 - \frac{z^{-1}}{\rho_k^R + j\rho_k^I}}, \qquad (10)$$

where ζ_k^R and ζ_k^I are the real and imaginary parts of the kth zero, while ρ_k^R and ρ_k^I are the real and imaginary parts of the kth pole. If a root (a pole or zero) is real, the imaginary part is specified as zero. If a root is complex, its conjugate must also be present. If a root is zero, then the term associated with it is implemented as z^{-1}, rather than $(1 - z^{-1}/r)$ (where r is the root). If $\rho_k^R < 0$ then the kth pole is stable.

Example:

```
V(out) <+ zi_zp(V(in), {0, 0}, {–1, 0});
```

implements

$$\frac{V_{out}(z)}{V_{in}(z)} = \frac{z^{-1}}{1 + z^{-1}}. \qquad (11)$$

zi_nd implements the rational polynomial form of the z transform filter. The general form is,

zi_nd(*operand*, *n*, *d*, *T*, <τ>, <t_0>);

where n is a vector of M real numbers containing the coefficients of the numerator and d is a vector of N real numbers containing the coefficients of the denominator. The transfer function is

$$H(z) = \frac{\sum_{k=0}^{M} n_k z^{-k}}{\sum_{k=0}^{N} d_k z^{-k}}. \tag{12}$$

Example:

```
V(out) <+ zi_nd(V(in), {1}, {1}, 50u, 10u, 200u);
```

This example implements a simple sample and hold. When driven with a 1 kHz 1V sine wave it produces the output shown in Figure 4.

FIGURE 4 *Example input and output waveform for a z-domain sample-and-hold.*

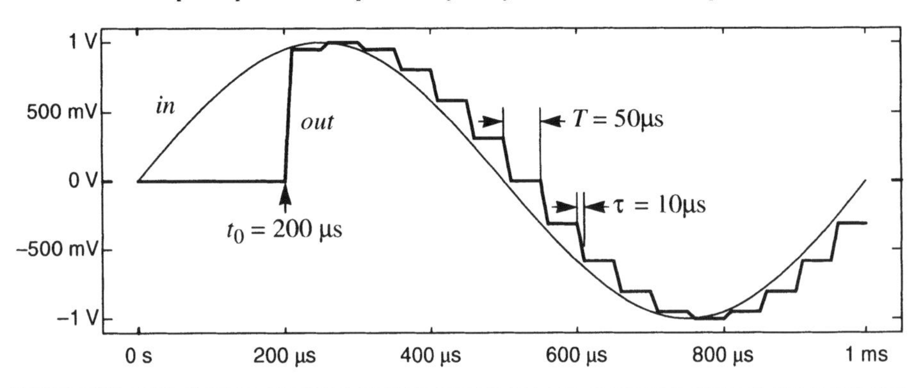

Example:

```
V(out) <+ zi_nd(V(in), {1}, {0, –1});
```

implements

$$\frac{V_{out}(s)}{V_{in}(s)} = \frac{1}{1 - z^{-1}}, \tag{13}$$

which is a discrete-time integrator.

The filters *zi_zd* and *zi_np* are similar to the z transform filters already described with *zi_zd* accepting a zero/denominator polynomial form and *zi_np* taking a numerator polynomial/pole form.

zi_zd(*operand*, ζ, d, T, <τ>, <t_0>)
zi_np(*operand*, n, ρ, T, <τ>, <t_0>);

Difference Equations and Z-Domain Filters. The z filters are used to implement the equivalent of discrete-time filters on continuous-time signals. Discrete-time filters are characterized as being either finite-impulse response (FIR) or infinite-impulse response (IIR). These filters are often defined in terms of difference equations. For example,

$$y_n = \sum_{k=1}^{N} a_k y_{n-k} + \sum_{k=0}^{M} b_k x_{n-k} \tag{14}$$

is a difference equation that describes an FIR filter if $a_k = 0$ for all k and an IIR filter otherwise. A block diagram that implements this filter is shown in Figure 5. In this diagram, the blocks marked "z^{-1}" represent unit delay cells; they delay the signal that passes through them by one sample interval. The transfer function of this block diagram is given by

$$H(z) = \frac{\sum_{k=0}^{M} b_k z^{-k}}{1 - \sum_{k=1}^{N} a_k z^{-k}}, \tag{15}$$

which can be implemented with a *zi_nd* filter if $n_k = b_k$ for all k, $d_1 = 1$ and $d_k = -a_k$ for $k > 1$.

FIGURE 5 *Block diagram representation of an nth-order difference equation.*

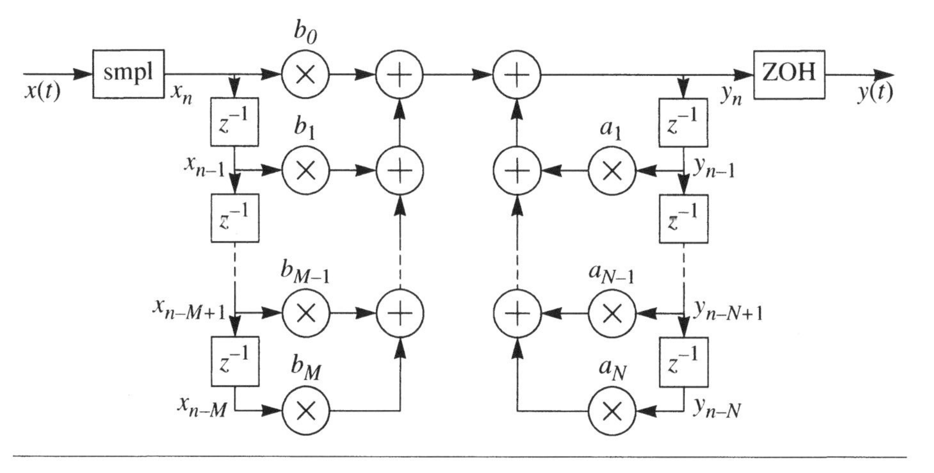

4.7 Thresholding Functions

These functions detect and return information about threshold crossings. They are subject to the same restrictions as analog operators (Section 4.6).

4.7.1 Cross and Above

The *cross* and *above* functions produce events when they detect a threshold crossing. They are described in Section 6.8 on page 204.

4.7.2 Last Crossing

The *last_crossing* function returns a real value representing the time in seconds when its operand last crossed zero in a specified direction. The general form is

last_crossing(*operand, direction* **);**

If *direction* is +1 the function will observe only rising transitions through zero; if −1, falling transitions are observed; if 0, both rising and falling transitions are observed, and if any other value, no transitions are observed.

The *last_crossing* function does not control the time step to get accurate results; it uses interpolation to estimate the time of the last crossing. However, it can be used with the *cross* function for improved accuracy. See Listing 25 on page 82 for an example of this.

4.8 Limiting Functions

4.8.1 Limited Exponential

The *limexp* function is an operator whose internal state contains information about the argument on previous iterations[†]. It returns a real value that is the exponential of its single real argument, however, it internally limits the change of its output from iteration to iteration in order to reduce the risk of overflow and improve convergence. On any iteration where the change in the output of the *limexp* function is bounded, the simulator is prevented from terminating the iteration process. Thus, the simulator can only converge when the output of *limexp* equals the exponential of the input. The general form is

limexp(*x* **)**

† Unlike the analog operators, the *limexp* function does not retain information about previous time points. Rather, it retains information about the previous iterations at the current time point.

and the return value is e^x. The apparent behavior of *limexp* is not distinguishable from *exp*, except using *limexp* to model semiconductor junctions generally results in dramatically improved convergence, though at the cost of extra memory being required.

4.9 Small-Signal Stimulus Functions

A small-signal analysis computes the steady-state response of a system that has been linearized about its operating point and is driven by one or more small sinusoids. The process of linearization eliminates the possibility of driving the circuit with conventional behavioral statements. The small-signal stimulus functions are provided to address this need; they operate after the linearization. They are demonstrated in Listing 16 on page 65 in Chapter 3.

4.9.1 AC Stimulus (ac_stim)

SPICE-class simulators provide AC analysis, which is a small-signal analysis used for computing transfer functions. Verilog-A/MS provides the *ac_stim* function as a way of providing the stimulus for an AC analysis.

ac_stim(*<analysisName>*, *<mag>*, *<phase>*)

The AC stimulus function returns zero during large-signal analyses (such as DC and transient) as well as on all small-signal analyses using names that do not match *analysisName*. The name of a small-signal analysis is implementation dependent, although the expected name (of the equivalent of a SPICE AC analysis) is 'ac', which is the default value of *analysisName*. When the name of the small-signal analysis matches *analysisName*, the source becomes active and models a source with magnitude *mag* and phase *phase*. The default magnitude is one and the default phase is zero and is given in radians.

4.9.2 Noise Stimulus (white_noise, flicker_noise, noise_table)

White noise processes are stochastic processes whose current value is completely uncorrelated with any previous or future values. This implies their spectral density does not depend on frequency. They are modeled using

white_noise(*pwr*, *<name>*)

which generates white noise with a power of *pwr*. The *white_noise* function can be used to model the thermal noise produced by a resistor as follows,

V(res) <+ r*I(res) + **white_noise**(4*`P_K*$temperature*r, "thermal");

The *flicker_noise* function models flicker noise. The general form is

flicker_noise(*pwr*, *<exp>*, *<name>*)

It produces noise with a power of *pwr* at 1 Hz and varies in proportion to $1/f^{\alpha}$ where α is given by the *exp* parameter. The default value for *exp* is 1, which corresponds to pink noise (noise whose power is proportional to $1/f$).

The *noise_table* function produces noise whose spectral density varies as a piecewise linear function of frequency. The general form is

noise_table(*vector*, <*name*>)

where *vector* contains pairs of real numbers: the first number in each pair is the frequency in Hertz and the second is the power. Noise pairs are specified in the order of ascending frequencies. The *noise_table* function performs piecewise linear interpolation to compute the power spectral density generated by the function at each frequency.

Each of the noise stimulus functions support an optional *name* argument, which acts as a label for the noise source. It is used when the simulator outputs a report that details the individual contribution made by each noise source to the total output noise. The contributions of noise sources with the same *name* from the same instance of a module are combined in the noise contribution summary.

4.10 User-Defined Functions

Verilog-A/MS provides the ability to define functions within a module. There are two types of user-defined functions, analog and digital. Analog functions are called from within an analog process and are described in Section 6.7 on page 204. Digital functions can be called from digital processes and tasks, or from continuous assignments. They are described in Section 7.9.1 on page 221.

5 System Functions and Tasks

This section describes the various functions that provide an interface to the internal algorithms of the simulator or to the operating system.

5.1 Simulator Interface

5.1.1 Bound Step ($bound_step)

The *$bound_step* function places a bound on the size of the next time step used by the continuous time kernel. It does not specify what the time step is, but rather how large it can be.

$bound_step(*maxStep*)

The maximum allowed time step is specified as the argument. There is no return value.

5.1.2 Discontinuity ($discontinuity)

The *$discontinuity* function is used to notify the continuous-time kernel of discontinuous changes in the behavior of an analog component. This information helps the kernel maintain accurate results in exceptional situations.

$discontinuity(*order*)

The argument is a non-negative integer that indicates the lowest order derivative that is discontinuous. An argument of *i* implies a discontinuity in the ith derivative of the constitutive equation with respect to either a signal value or time. Hence, *$discontinuity*(0) indicates a discontinuity in the model, *$discontinuity*(1) indicates a discontinuity in the model's derivatives, etc.

Discontinuous behavior can cause convergence problems for the simulator and should be avoided whenever possible. To this end, the filter functions (*transition*, *slew*, etc.) can be used to smooth behavior that is discontinuous with respect to time.

Discontinuities created by switch branches and built-in functions, such as *transition*, need not be announced.

5.1.3 Finish ($finish)

The *$finish* function terminates the simulation session. It takes an integer argument that indicates the message that should be printed. If 0, nothing is printed, if 1, the time and location are printed (the default behavior), and if 2, time and location are printed along with memory and CPU statistics.

$finish<(*n*)>

5.1.4 Stop ($stop)

The *$stop* function is similar to *$finish*, except that it causes the simulation to be suspended rather than terminated. The difference being that the user may resume simulation after it has been suspended, but cannot do so after it has been terminated.

$stop<(*n*)>

5.2 Display Tasks

The display tasks are used to produce textual output that is either sent to the display or to a file. They are all very similar in use, and so the *strobe* function is described in detail, and then it is contrasted against the other tasks.

5.2.1 Strobe ($strobe)

The $*strobe* task takes a list of arguments and converts them into text that is written to the standard output. The first argument must be a string, and that string may include special character patterns that incorporate subsequent arguments. Those characters are given in Table 4 on page 156 and Table 13. Once the arguments associated with the first string are exhausted, if there are any arguments remaining, the first must be a string, in which case the process repeats. Once the arguments have been printed, a newline is added. If there are no arguments, only the newline is printed.

TABLE 13 *Argument formatting codes for the display tasks.*

Code	Args	Result
%b, %B	1	Interpolate argument with binary format.
%c, %C	1	Interpolate argument with ASCII character format.
%d, %D	1	Interpolate argument with decimal format.
%e, %E	1	Interpolate real argument with exponential format.
%f, %F	1	Interpolate real argument with decimal format.
%g, %G	1	Interpolate real argument with exponential or decimal format.
%h, %H	1	Interpolate argument with hexadecimal format.
%m, %M	0	Display hierarchical name.
%o, %O	1	Interpolate argument with octal format.
%s, %S	1	Interpolate argument as a string.
%%	0	Display '%'.

Example:

```
$strobe("Average period = %g measured from %d periods.", period, total);
```

The format codes of Table 13 that act on real valued arguments support C language style format specifications. Optional integers can be inserted into the format code to specify the field width, the precision, or both. For example, the format specification '%10.3g' indicates the number should be formatted with a minimum field width of 10 and with 3 digits in the mantissa.

5.2.2 Display ($display)

The $*display* task is identical to the $*strobe* task.

5.2.3 Write ($write)

The $*write* task is identical to the $*strobe* task, except newlines are not added unless explicitly specified in the format strings.

5.2.4 Monitor ($monitor)

The $*monitor* task is identical to $*strobe*, except output is only produced when the value of the arguments change.

5.3 File Operation Tasks

Verilog maintains a table of open files that may contain at most 32 files. Each file corresponds to one bit in a 32 bit integer that is referred to as a multichannel descriptor. The first bit, or channel 0, corresponds to the standard output. The first call to *fopen* opens channel 1, which corresponds to the second bit, etc. This non-traditional approach to files allows output to multiple files with a single statement.

5.3.1 File Open ($fopen)

The $*fopen* task takes a string argument that is interpreted as a file name and opens the corresponding file for writing. It returns an integer that contains the multichannel descriptor for the file. A 0 is returned if the file could not be opened for writing.

$fopen(*filename*)

5.3.2 File Close ($fclose)

The $*fclose* task takes an integer argument that is interpreted as a multichannel descriptor for a file or files. It closes those files and makes the channels that were associated with the files available for reuse.

$fclose(*multichannelDescriptor*)

5.3.3 Display to File ($fdisplay)

The $*fdisplay* task is identical to the $*display* task, except that it writes to one or more files rather than to the standard output. The files are specified in the form of a multichannel descriptor that is given as the first argument.

$fdisplay(*multichannelDescriptor*, *formatString*, <...>)

5.3.4 Strobe to File ($fstrobe)

The *$fstrobe* task is identical to the *$strobe* task, except that it writes to one or more files rather than to the standard output. The files are specified in the form of a multi-channel descriptor that is given as the first argument.

$fstrobe(*multichannelDescriptor, formatString*, <...>)

5.4 Random Numbers

These functions return a number chosen at random from a random process with a specified distribution. When called repeatedly, they return a sequence of random numbers. Each takes an inout argument, named *seed*, that specifies the sequence. A different initial seed results in a different sequence. The seed must be a simple integer variable that is initialized to the desired initial value. This variable is updated by the function on each call.

5.4.1 Uniformly Distributed Integers ($random)

The *$random* function returns a randomly chosen 32 bit integer. Each time it is called it returns a different value with the values being distributed uniformly over the range of 32 bit integers.

$random<(*seed*)>

5.4.2 Uniformly Distributed Real Numbers ($rdist_uniform)

The *$rdist_uniform* function returns a randomly chosen real number that falls in a specified interval. The interval is specified by two real valued arguments that give the lower and upper bound of the interval.

$rdist_uniform(*seed, lowerBound, upperBound*)

5.4.3 Normally Distributed Real Numbers ($rdist_normal)

The *$rdist_normal* function returns a randomly chosen real number from a population that has a normal (Gaussian) distribution. The distribution is parameterized by its mean and its standard deviation.

$rdist_normal(*seed, mean, standardDeviation*)

5.4.4 Exponentially Distributed Real Numbers ($rdist_exponential)

The *$rdist_exponential* function returns a randomly chosen real number from a population that has an exponential distribution. The distribution is parameterized by its mean.

$rdist_exponential(*seed*, *mean*)

5.4.5 Poisson Distributed Real Numbers ($rdist_poisson)

The *$rdist_poisson* function returns a randomly chosen real number from a population that has a Poisson distribution. The distribution is parameterized by its mean.

$rdist_poisson(*seed*, *mean*)

5.4.6 Chi-Squared Distributed Real Numbers ($rdist_chi_square)

The *$rdist_chi_square* function returns a randomly chosen real number from a population that has a Chi Square distribution. The distribution is parameterized by the degrees of freedom (must be greater than zero).

$rdist_chi_square(*seed*, *degreeOfFreedom*)

5.4.7 Student T Distributed Real Numbers ($rdist_t)

The *$rdist_t* function returns a randomly chosen real number from a population that has a Student T distribution. The distribution is parameterized by the degrees of freedom (must be greater than zero).

$rdist_t(*seed*, *degreeOfFreedom*)

5.4.8 Erlang Distributed Real Numbers ($rdist_erlang)

The *$rdist_erlang* function returns a randomly chosen real number from a population that has an Erlang distribution. The Erlang distribution describes the time spent waiting for k Poisson distributed events. The distribution is parameterized by its mean and by k (must be greater than zero).

$rdist_erlang(*seed*, *k*, *mean*)

6 Analog Behavior

This section describes the facilities that are available in Verilog-A/MS for describing analog behavior. As such, the description for this behavior would be found in an analog process. Everything in this section is presented with the assumption that what is being described is found within an analog process. Behavioral description found outside analog processes is described in Section 7 starting on page 208.

6.1 Analog Processes

The basic building block of a behavioral model is the ***process***. A process is an independent thread of control. A system generally consists of many processes, all of which operate concurrently and interact. Processes represent an essential difference between hardware description languages and general-purpose programming languages such as C or C++. A process might be very simple, involving only one action or behavior, perhaps applied repeatedly or continuously, or it might involve a complex algorithm. The art of modeling hardware is to conceive of the behavior of systems as a set of these independent but communicating processes.

Discrete time behavioral definitions are encapsulated within the *initial* and *always* processes (5§7.1p209). For continuous time simulation, the behavioral description is encapsulated within an analog process. Analog processes are introduced using the *analog* keyword.

analog *statement*;

Only one analog process is allowed in a module definition.

Example:

```
analog
    V(out) <+ r*I(out);
```

An analog process is evaluated 'at all time' by the continuous-time kernel. Practically this means that every analog process is executed at every time point, and while it executes, time is frozen at a particular instant of time as chosen by the simulator.

Analog processes do not support the concept of pausing or blocking as described in Section 7.6 on page 216. As such, delay (using '#') and *wait* statements are not allowed. In addition, as described in Section 6.8, the meaning of event statements (using '@') has been modified within analog processes to eliminate blocking.

6.2 Procedural Blocks

Notice that the analog process consists of a single statement, which is only sufficient for simple behavior. For more complex behavior, a procedural block should be used. A ***procedural block*** is defined as a sequence of statements surrounded by a *begin-end* pair. As such, a procedural block is also sometimes referred to as a sequential block. When the block is executed, the statements are executed in order. Anywhere a single statement is allowed, a procedural block can be used instead.

```
begin
    statement;
    statement;
    ...;
end
```

Examples:

```
analog begin
    I(res) <+ is∗(limexp(V(res)/$vt) – 1);
    qd = tf∗I(res) – 2∗cjo∗phi∗sqrt(1 – V(cap)/phi);
    I(cap) <+ ddt(qd);
end

analog begin
    @(above(V(n) – threshold)) begin
        if (!given) begin
            $strobe(message);
            given = 1;
        end
    end
end
```

In the first example, a procedural block was used to allow an analog process to be described with more than one statement. Within the block, the statements are executed in order. In the second example, three nested procedural blocks are used, showing that this idea of replacing a single statement with a sequence of statements is useful in many places.

6.2.1 Named Blocks

One can name a procedural block by adding a colon and a name after the *begin* keyword. Doing so makes it possible to declare local variables within the block, as demonstrated in the example below. Notice that in this code fragment, which is a modification of the one above, the variable *given* is declared within the named block *fault*, which is associated with the @ statement.

Example:

```
analog begin
    @(above(V(n) – threshold)) begin : fault
        integer given;
```

```
            if (!given) begin
                $strobe(message);
                given = 1;
            end
        end
    end
```

To understand this code, one must remember that all variables declared within modules, even those declared in named blocks, retain their values for the entire duration of the simulation. Thus, even when the flow of execution leaves the context where the variable *given* is declared, its value is retained and available when the flow of execution returns.

6.3 Assignments

While branch contributions and indirect branch assignments are used for modifying signals, procedural assignments are used for modifying integer and real variables. Assignment statements consist of a variable (the target) followed by an assignment operator (=) followed by an expression. The value of the variable is replaced by the value of the expression. When the statement is executed, the expression is evaluated and the target updated before proceeding to the next statement.

Example:

```
qd = tf*I(res) – 2*cjo*phi*sqrt(1 – V(cap)/phi);
```

Any variable or register that is assigned a value from within an analog process is captured by that process, meaning that it is not possible to assign it a value outside that process. Conversely, if a variable is assigned a value outside an analog process, say in a initial or always process, it is not possible to also assign it a value from within an analog process.

6.4 Contributions

Contribution statements (also described in Section 3.2 on page 169 and in Chapter 3 in Section 4 starting on page 58) are used to modify the signals on branches. Contribution statements consist of a signal (an access function applied to a branch) followed by a contribution operator (<+) followed by an expression. The first contribution to a particular signal within a module sets the value of the signal. Subsequent contributions add to that value.

Example:

```
analog begin
    V(p,n) <+ 2*dc;
    V(p,n) <+ r*I(p,n);
    V(p,n) <+ 2*ac_stim(, mag);
    V(p,n) <+ white_noise(4*`P_K*$temperature*r, "thermal");
end
```

In this example, which is paraphrased from Listing 16 on page 65, the first contribution statement contributes a constant value giving the branch the attributes of a large-signal DC source. The second contributes a value that varies in proportion to the current through the branch, giving the branch the attributes of a resistor. Since this contribution adds to the first, the branch behaves as a resistive DC source. The third contribution layers on the attributes of an small-signal AC source, and the fourth supplies the attributes of a noise source. Together they describe a noisy resistive branch that acts as a source in both large- and small-signal analyses. An important thing to notice about this example is that the contribution statements can be given in any order. This is a direct consequence of the commutative property of addition.

Contribution statements provide the ability to simultaneously determine the values of multiple mutually dependent unknowns. For example, with a resistor

```
I(p,n) <+ V(p,n)/r;
```

the contribution statement provides a relationship between the branch current and the branch voltage that, along with other such relationships that may be found both within the same module and in other modules, will be solved to determine the values that simultaneously satisfy all of the relationships. This ability to solve simultaneous equations distinguishes the contribution statement from simple procedural assignment. And because contribution statements can only exist within *analog* processes, it generally determines when an *analog* process is needed and, conversely, when one can instead use the generally faster event driven processes (*initial* or *always*).

Consider three examples. First a conservative model, such as a resistor, embedded in a arbitrary circuit. Here, the voltage impressed on the resistor by the circuit will be a function of the current being supplied to the resistor. Thus, we cannot know the current through the resistor without knowing its voltage, but we also cannot know its voltage without knowing its current. These two quantities must be determined by solving for all of the unknowns simultaneously, meaning that the resistor model (as well as other models in the circuit) must contain an analog process that further contains one or more contribution statements.

Now consider a signal-flow model such as a voltage amplifier with an arbitrary external signal-flow feedback network. Again, one cannot know the output of the amplifier without knowing the input voltage, but the input voltage is possibly a function of the current value of the output voltage as processed by the feedback network. So the amplifier (and the modules within the feedback network) must contain an analog process that further contains one or more contribution statements. This is true even if the signals being processed by the amplifier are discrete-event by nature (piecewise constant).

Finally, consider a signal-flow model such as the amplifier of the last example, and assume both that it is processing discrete-event signals and that any feedback that is present includes sufficient delay so that the values fed back are previously known (they are from a previously computed solution point). In this case, one can always directly compute the output from the known quantities at the input, and so contribution statements are not required. Furthermore, the signals being processed are discrete-event signals and so an analog process is not needed. In this case, one can successfully use an event-driven process such as an *always process*.

6.5 Conditionals

6.5.1 Conditional Operator

The ?: operator is provided as a way of performing conditional operations within an expression. It takes three arguments in the following form

cond ? *val1* : *val2*

where *cond*, *val1*, *val2*, and the combination, are all expressions. The operator returns *val1* if *cond* is nonzero, otherwise it returns *val2*.

Example:

```
state = (V(d) > 0) ? 1 : -1;
```

In this example, *state* becomes 1 if *V*(*d*) is greater than 0 and –1 otherwise.

6.5.2 If-Else Statement

The *if-else* statement conditionally evaluates statements based on the value of a logical expression. If the expression evaluates to true (that is, has a non-zero value) the *if clause* is executed and if the *else clause* exists, it is not evaluated. If it evaluates to false (has a zero value) or if the result is ambiguous (has a value of *x* or *z*), the *if clause* is not executed and if the *else clause* exists it is evaluated.

if (*logical expression*)
    if clause;
[**else**
    else clause;]

Examples:

```
if (V(ps,ns) > thresh)
    V(p,n) <+ 0;
else
    I(p,n) <+ 0;
```

and

```
if (V(ps,ns) > thresh)
    V(p,n) <+ 0;
```

If-else statements may not contain analog operators unless the logical condition is a *genvar* expression.

6.5.3 Case Statements

The *case statement* is a multi-way decision statement that tests if an expression matches one of a number of other expressions, and if so, branches accordingly.

Example:

```
case (select)
    0: out = V(in0);
    1: out = V(in1);
    2: out = V(in2);
    3: out = V(in3);
    default: out = 0;
endcase
```

The case item expressions are evaluated and compared in the exact order that they are given. During this linear search, if one of the case item expressions matches the case expression given in parentheses, then the statement associated with that case item is executed. If all comparisons fail, and the default item is given, then the default item statement is executed; otherwise none of the case item statements are executed.

The case expression and the case item expressions are evaluated at runtime; neither is required to be a constant expression.

Case statements may not contain analog operators unless the case expression and the case item expressions are all *genvar* expressions.

There are also *casex* and the *casez* versions of the case statement. They operate on bit vectors: *casex* ignores bit positions that contain *x* or *z*, and *casez* ignores bit positions that contain *z*. They both use '?' as don't cares in bit patterns.

6.6 Iterators

There are several types of looping statements: *repeat*, *while*, and *for*. These statements provide a means of evaluating a statement a specified number of times.

6.6.1 Repeat and While Loops

A *repeat* loop executes a statement a fixed number of times. Evaluation of the expression determines how many times a statement is executed.

Example:

```
repeat (size) begin
    memory[i] = 0;
    i = i + 1;
end
```

A *while* loop executes a statement until an expression becomes false. If the expression starts out false, the statement is not executed at all.

```
while (temp) begin
    if (temp[0])
        counter = counter + 1;
    temp = temp >> 1;
end
```

Analog operators and contribution statements are not allowed in the *repeat* and *while* iterators.

6.6.2 For Loops

The *for* statement is a flexible looping construct taken from the C programming language.

```
for (i = bits – 1; i >= 0; i = i – 1) begin
    if (V(in[i]) > thresh)
        aout = aout + fullscale/weight;
    weight = weight*2;
end
```

Operation of the *for* statement is controlled by the three statements that are contained within the parentheses that follow the *for* keyword. The first statement is evaluated once before the loop is entered; it is used as an initializer. The second is a logical expression that is evaluated before each iteration; if the value is true the iteration proceeds, if it is false the loop terminates without executing the iteration. The third is evaluated at the end of each iteration; it is generally used to update the index variable. Executing the iteration involves executing the statement that follows the *for* statement, in this case a composite statement that is contained between the *begin* and the *end* keywords. The above *for* loop is equivalent to the following while loop.

```
i = bits – 1;
while (i >= 0) begin
    if (V(in[i]) > thresh)
        aout = aout + fullscale/weight;
    weight = weight*2;
    i = i – 1;
end
```

For statements may not contain analog operators or contribution statements unless the index variable is a *genvar.*

6.6.3 Generate Loops (Deprecated)

The *generate* statement is an obsolete looping construct that was intended to allow support for looping with restricted constructs such as analog operators. The *for* loop that uses a *genvar* index is intended to replace the *generate* statement. A generate loop has the form

```
generate index (start, end, <increment>)
    body
```

The generate statement takes an integer index, a pair of bounds, and an optional increment. The start and end bounds and the increment are constant expressions. They are only evaluated at elaboration time. If the expressions used for the increment and bounds change during the simulation, it does not affect the behavior of the generate statement. In effect, the generate loop is unrolled in advance.

Example:

```
generate i (bits–1, 0, 1) begin
    V(out[i]) <+ transition(result[i], td, tt);
end
```

In this example, if *bits* is assumed to be 4, the loop is equivalent to

```
V(out[3]) <+ transition(result[3], td, tt);
V(out[2]) <+ transition(result[2], td, tt);
V(out[1]) <+ transition(result[1], td, tt);
V(out[0]) <+ transition(result[0], td, tt);
```

The index must not be assigned or modified in any way within the loop.

If the lower bound is less than the upper bound and the increment is negative, or if the lower bound is greater than the upper bound and the increment is positive, then the generate statement does not execute. If the lower bound equals the upper bound, the increment is ignored and the statement executes once. If the increment is not given, it is taken to be +1 if the lower bound is less than the upper bound, and –1 if the lower bound is greater than the upper bound.

6.7 User-Defined Analog Functions

A function is code that is encapsulated and parameterized so that it can be shared throughout a module. An analog function definition begins with the keywords *analog function*, optionally followed by the type of the return value of the function, then the name of the function and a semicolon. One or more input parameters may be declared along with any number of variables. The body of the function consists of a single statement that follows the declarations, and the function ends with the keyword *endfunction*. The return type may be *real* or an *integer*, with the default being *real*. The return value is set by assigning a value to a variable whose name is the same as the name of the function.

An analog function:

- may contain any statements available for conditional execution;
- must not contain access functions, contribution statements, or event statements;
- must have at least one input declared;
- must not contain named blocks; and
- may only reference locally-defined variables or variables passed as arguments.

Example:

```
analog function real sinc;
    input arg;
    real arg;

    begin
        if (arg != 0)
            sinc = sin(arg)/arg;
        else
            sinc = 1;
    end
endfunction
```

6.8 Analog Events

An event is an occurrence of a particular change in the state of the circuit. They are detected by setting up a statement that looks for the desired change. When the event occurs, an action is taken. Thus, analog event statements consist of two parts, the part that specifies the event and the part that specifies the action to be taken when the event occurs. This is different from event statements found in initial and always processes, which simply block execution of the process until the event occurs (5§7.6.2p216).

```
@(event-expression)
    action;
```

Unlike event statements in *initial* and *always* processes, event statements in analog processes are non-blocking. Thus, execution does not wait on an event in an analog process, it simply passes over the event statement except at the instant the event occurs. In other words, at the instant the event occurs the action is executed. At other times, the event statement is simply bypassed.

Example:

```
@(cross(V(clk) – vth, dir))
    state = (V(d) > vth);
```

The action of an event statement must not contain an analog operator or a contribution statement.

An event expression consists of one or more event monitors separated by the *or* operator. The event action is triggered if an event is produced by any of the monitors. The following sections describe the available event monitors.

In Verilog-AMS event expressions found in analog processes may contain the traditional Verilog event expressions described in Section 7.6 on page 216. In this way, events that occur in registers and event-driven nets can affect the behavior of analog processes. Similarly, the event expressions described in this section can be used in initial and always processes.

6.8.1 Initial and Final Step

The *initial_step* and *final_step* events occur on the first and last point of a particular simulation. Both accept optional arguments that specify the particular analyses (or type of simulation) that must be running for the events to occur. The analyses are given in the form of quoted strings. Which strings are associated with particular analyses is implementation dependent, however “ac”, “dc”, “noise” and “tran” are used for the traditional SPICE analyses (see Table 11 on page 177).

Examples:

```
@(final_step) begin
    if (faults)
        $strobe(“%d faults occurred in %m.\n”, faults);
end
```

and

```
@(final_step("tran")) begin
    if (count)
        $strobe("average delay = %g.\n", tsum/count);
end
```

6.8.2 Autonomous Events

The *timer* function produces analog events at specific points in time. The general form is

timer (*startTime*, *<period>*, *<timeTol>*)

where *startTime* is required; *period* and *timeTol* are optional arguments. All arguments are real expressions.

The *timer* function schedules an event that occurs at, or just beyond, *startTime*. The event is guaranteed to be within *timeTol* of the requested time. If the *period* is specified and is greater than zero, the timer function schedules subsequent events at multiples of *period* from *startTime*.

An example of the *timer* event function is shown in the pseudo-random bit stream generator given in Listing 2.

LISTING 2 *Pseudo-random bit stream generator.*

```
module bitStream (out);
    parameter real start = 0.0;                    // time that the first bit is emitted (s)
    parameter real period = 1.0 from (0:inf);      // period of bit updates (s)
    output out;
    electrical out;
    integer x;

    analog begin
        @(timer(start, period))
            x = ($random >= 0) ? 1 : 0;
        V(out) <+ transition( x, 0.0, period/100.0 );
    end
endmodule
```

6.8.3 Threshold Crossings

Two functions are available that monitor threshold crossings, *cross* and *above*. The *above* function is not yet accepted as part of the standard and so may not be in all implementations.

The *cross* function is used for generating an analog event when the result of an expression passes through zero in a particular direction. In addition, *cross* controls the time step to accurately resolve the crossing. The general form is

cross (*argument*, *<direction>*, *<timeTol>*, *<exprTol>*)

where *argument* is required, and *direction*, *timeTol*, and *exprTol* are optional. All arguments are real expressions, except *direction*, which is an integer expression.

If the direction is specified as 0 or is not specified, the event and time step control occur on both positive and negative going zero crossings of the *argument*. If *dir* is +1 (or –1), the event only occurs on rising (falling) transitions of the signal. If any other value is specified, the *cross* function is disabled.

The underlying implementation will have limited precision and so the event may not occur precisely at the threshold crossing. Tolerances can be specified to control the accuracy of the timing of the event. The event will always occur at, or just beyond, the zero crossing. If *exprTol* is given then the event will occur while the absolute value of the argument is less than *exprTol*. If *timeTol* is specified, then the event will occur within *timeTol* seconds of the zero crossing.

An application of the *cross* event function is given in the D-flip-flop of Listing 3.

LISTING 3 *D flip flop.*

```
module dff (q, d, clk);
    parameter integer dir = +1 from [–1:+1] exclude 0;
            // if dir=+1, rising clock edge triggers flip flop
            // if dir=–1, falling clock edge triggers flip flop
    output q; voltage q;            // Q output
    input clk; voltage clk;         // Clock input (edge triggered)
    input d; voltage d;             // D input
    integer state;

    analog begin
        @(cross(V(clk) – 0.5, dir))
            state = (V(d) > 0.5);
        V(q) <+ transition( state ? 1 : 0 );
    end
endmodule
```

The *above* function is similar to the *cross* function, except that while *cross* only produces events in an analysis that advances time, *above* will also fire in equilibrium point analyses (such as a DC operating point or a swept DC analysis). It does not accept a direction specifier, instead it generates an event as the argument becomes positive. The general form is

above(*argument*, *<timeTol>*, *<exprTol>*)

It produces events

- When the argument transitions from negative to positive during a swept analysis (swept DC) or an analysis where time is advancing (transient, etc.), or
- during a DC operating point analysis, an IC analysis, or the first point of a DC sweep if the argument is positive.

An application of the *above* event function is shown in Listing 4. It is sentinel code that produces a warning message as a breakdown voltage is exceeded.

LISTING 4 *Warn on breakdown.*

```
module breakdown (n);
    parameter real threshold = 1000.0;        // message emitted when this level exceeded (V)
    parameter message = "Warning: breakdown voltage exceeded.";
                                              // the message to print
    input n; voltage n;
    integer given;

    analog begin
        @(above(V(n) – threshold)) begin
            if (!given)
                $strobe(message);
            given = 1;
        end
    end
endmodule
```

7 Discrete-Event Behavior

The analog process (5§6.1p196) is the only construct that can be used to hold behavior that has primarily continuous time semantics. For discrete-event behavior, there are several different constructs available. For simple behavior, continuous assignments and net assignments can be used. They are described in Sections 7.4.2 and 7.4.3. For more complex behavior, *initial* and *always* processes are used.

7.1 Initial and Always Processes

Every ***initial*** and ***always process*** starts a separate concurrent activity flow as simulation begins (4§2.1.4p103). The flow takes the form of a statement or a block of statements. The initial process will run through that flow once and then terminate. The always process continually repeats its flow, never exiting or stopping. It is an endless loop. However, execution of the flow can pause at particular points, waiting either for a particular interval of time to expire or for some change to occur in an external process that allows execution to continue. The language constructs that temporarily block the execution of the process are described in Section 7.6. In this way, the initial and always processes are different from the analog process in a fundamental way. An initial process often blocks and an always process should always block, but analog processes never block. The semantics of statements that are allowed in analog processes do not support it.

Initial and always processes are evaluated by the discrete-event kernel. They take the form,

initial *statement*;
always *statement*;

Example:

```
module clock (out);
    output out;
    reg out;

    initial out = 0;

    always #50 out = ~out;
endmodule
```

In this example execution of both the initial and always processes starts simultaneously. The initial process executes its one statement and then terminates. It is used to initialize the value of *out*. The always process blocks (pauses) for 50 time units and then executes a statement that inverts the contents of the *out* register. It then restarts the process, and will do so repeatedly until the simulation is terminated. In this way, this module produces an output that toggles between 0 and 1 and back with a period of 100 time units.

7.2 Procedural Blocks

As with analog processes (5§6.2p196), procedural blocks are used with initial and always processes to describe more complex behavior. A ***procedural block*** is defined as a sequence of statements surrounded by a *begin-end* pair (4§2.1.7p106). As such, a procedural block is also sometimes referred to as a sequential block. When the block is executed, the statements are executed in order. Anywhere a single statement is allowed, a block can be used instead.

```
begin
    statement;
    statement;
    ... ;
end
```

Example:

```
module decade_counter (out, clk);
    input clk;
    output [3:0] out;
    reg [3:0] out;

    initial out = 0;

    always begin
        @(posedge clk)
            out = out + 1;
        if (out > 10)
            out = 0;
    end
endmodule
```

In this example, a procedural block allows an always process to be described with more than one statement. Within the block, the statements are executed in order.

7.2.1 Named Blocks

As mentioned before, procedural blocks can be named by adding a colon and a name after the *begin* keyword (5§6.2.1p197). Naming a block creates a new scoping level, making it possible to declare local variables within the block. It also makes it possible to disable a block. When a block is disabled, it terminates execution of the block and begins execution at the statement that follows the block.

```
disable block_name;
```

It is not possible to disable a named block that is contained within an analog process, but it is possible to disable a block running in a remote process (by using hierarchical names (5§9.4p230)) as well as disable the current block, as shown in this example.

Example:

```
begin : break
    for (i = 0; i < n; i = i + 1) begin : continue
        if ( ... )
            disable continue;        // proceed to next iteration
        if ( ... )
            disable break;           // exit the for loop
        ...
    end
end
```

The disable statement is used to mimic the functionality of *break* and *continue* statements in C. When **disable** continue; is executed, it causes the *continue* block to be terminated, which has the effect of causing the for loop to proceed to the next iteration. When **disable** break; is executed, it causes the *break* block to be terminated, which has the effect of causing the for loop to exit completely.

7.3 Concurrent Blocks

An alternative to procedural blocks, which executes a group of statements sequentially, is concurrent blocks, which executes a group of statements in parallel. In other words, while a procedural block runs a group of statements by executing them in order and waiting for each to complete before starting the next, the ***concurrent block*** starts each statement simultaneously and waits for the last to complete before exiting the block (4§2.1.8p106). A concurrent block is a group of statements surrounded by a *fork-join* pair.

```
fork
    statement;
    statement;
    ... ;
join
```

Example:

```
module counter16 (out, clk, reset);
    input reset, clk;
    output [15:0] out;
    reg [15:0] out;
```

```
always fork : main
    out = 0;
    forever begin
        @(posedge clk)
            out = out + 1;
    end
    @(posedge reset)
        disable main;
join
endmodule
```

In this example, *fork* causes three statements to be launched simultaneously. The first initializes *out* to 0 and then completes. The second is a *forever* loop (5§7.8.1p220) that counts the number of rising edges on *clk*. Since it is a *forever* loop, it will never complete on its own. The last statement immediately blocks and waits for a rising edge on the *reset* input. As long as the edge does not come, the *forever* loop will continue to count. When the rising edge on the *reset* line occurs, the *main* concurrent block is disabled or terminated. At this point the always process cycles and restarts the *main* concurrent block, which simultaneously reruns the three statements, resetting the counter and restarting the count.

7.4 Assignments

7.4.1 Procedural Assignments

Procedural assignment statements function in a manner similar to assignment statements in traditional programming languages such as C and C++. They are found in initial and always processes.

target = expr;

They consist of a target, which is a variable name found on the left side of the assignment statement, and an expression, found on the right side. When the flow of control reaches the procedural assignment, the expression is evaluated. Then the value of the expression replaces the contents of the target variable. There are two types of procedural assignments that differ in when the target is updated. In the regular assignment, which uses '=' as an assignment operator, the targets is updated before beginning execution of the next statement.

Example:

```
out = ~out;
```

The regular assignment is sometimes referred to as a ***blocking assignment*** because it may include a construct that causes the assignment to be delayed, during which time execution of the process blocks (5§7.6.4p218).

A ***non-blocking procedural assignment*** is similar, but it uses '<=' as the assignment operator

target <= *expr*;

The actual update of the target occurs only after the process that contains the assignment blocks. And in fact, the update only occurs after all of the right-hand sides of every non-blocking assignment statement waiting on the same edge in the entire design have been evaluated. As a result, the new value of the target is not available in the next procedural statement (4§2.1.10p108).

Example:

```
@(posedge clk) begin
    a <= b;
    b <= a;
end
```

In this example, the values of *a* and *b* are swapped at every rising edge on *clk*. The semantics of the non-blocking assignment eliminate the need for an additional variable to temporarily hold the value of *a* and result in *a* and *b* being simultaneously updated. These same semantics can sometimes result in seemingly anomalous results when blocking and non-blocking assignment are combined.

Example:

```
a = 25;
a <= 3;
b = a;
```

In this example, *b* becomes 25 even though the second statement changes *a* to 3 because the value of *a* does not change immediately. Indeed, the flow of control does not stop (block) to update *a*, rather it keeps going until blocked by some other statement, and at that point *a* is updated.

7.4.2 Continuous Assignments

Like a regular procedural assignment, ***continuous assignments*** consist of a target followed by the '=' assignment operator and an expression, but in this case they all follow the *assign* keyword and the target must be a net rather than a variable (4§2.1.3p102).

assign *target* = *expr*;

Example:

```
assign a = b & c;
```

Continuous assignment statements are found outside the initial and always procedural blocks. As such, they are not evaluated at isolated points in time. Rather, they are continuously enforced. Effectively, any time any of the arguments in the expression, in this case *b* and *c*, change, the expression is re-evaluated and the target updated. They are used to behaviorally define combinational logic.

7.4.3 Net Assignments

Net assignments are simply the combination of continuous assignments with a net declaration.

Example:

```
wire c = a ^ ~b;
```

This is equivalent to

```
wire c;
assign c = a ^ ~b;
```

7.4.4 Procedural Continuous Assignments

Continuous assignments allow for the description of combinational logic whose output is updated any time any one of the inputs change. There is a procedural version of the continuous assignment statement that allows for continuous assignments to be made to registers for specified periods of time. To understand their use, consider the following model of a simple *d*-flip flop.

```
module dff (q, d, clk, clear);
    input clear, clk, d;
    output q;
    reg q;

    always @(posedge clk)
        q = #5 d;

    always @(posedge clear)
        #5 assign q = 0;

    always @(negedge clear)
        #5 deassign q;
endmodule
```

In this example, the first always process implements the behavior of the flip-flop relative to changes in the *clk* input. When a rising edge occurs on *clk*, the *d* input is sampled and then assigned to the *q* output after a delay of 5 units of time. The second always process implements the behavior of the flip-flop when *clear* is asserted. In this case, the *assign* statement is evaluated on the rising edge of *clear*. This represents an important difference between a continuous assignment and a procedural continuous assignment. The procedural continuous assignment is found within a procedural block (in this case an always process) and is executed at a specific point in time, when the control flow reaches it. When this occurs, the expression on the right-hand side of the procedural continuous assignment is attached to the target, in this case 0 is attached to *q*, resetting the flip-flop. If the right-hand side were an expression with terms whose values could change over time, *q* would be automatically updated whenever the value of the terms changed. As long as an expression is attached to a register by a procedural continuous assignment, any non-continuous procedural assignments made to that register are ignored. In this way, rising edges *clk* will not cause *q* to change from 0 as long as *clear* remains true. The final always process detaches the expression from *q* when the value of the *clear* input becomes false. This does not directly result in *q* changing from 0, but does remove the override on *q* so rising edges on *clk* allow *q* to take the value of *d*.

7.5 Nets and Registers

Nets and registers play a central role in describing behavior outside the analog process. They differ in that registers are assigned values from within processes whereas nets (see Table 8 on page 165) are assigned values either by structural models or using continuous assignments. Things are somewhat different in analog processes from which analog nets can be assigned values indirectly through an access function using contribution statements (5§6.4p198).

Either nets or registers can be used as inputs to initial and always processes, to continuous assignments, and to structural models. In fact, the input port of a structural module that expects a traditional Verilog wire type could be fed with an expression involving registers. For example, it is common to invert the value of a register before passing it into a gate using the '~' operator.

Example:

```
nand (c, ~a, ~b);
```

Finally, while either nets or registers can be used as output ports of a module, only nets can be used as input ports.

7.6 Timing Control

Timing control in Verilog behavioral descriptions is performed by blocking or pausing execution of a process for some period of time. There are three mechanisms in which this occurs: delays, events, and waits.

7.6.1 Delays

Delay is added to an initial or always process using a '#' followed by a constant valued expression (made up of numbers and parameter values) that specifies the delay.

#const_expr statement;

This is shown in the following example, which was taken from the clock module given on page 209.

Example:

```
always #50 out = ~out;
```

The delay is added by the use of #50. It indicates that execution should be paused for 50 time units (as specified by the *`timescale* directive (5§1.4p151)). Notice that there is no semicolon between the delay specification and the subsequent statement. If no subsequent statement is desired, a semicolon should immediately follow the delay.

In addition to the delay described here, other forms of discrete-event delay are available. It is possible to associate delay with a net (5§2.5p164), a gate (5§9.3p229), and with an assignment statement (5§7.6.4p218).

7.6.2 Events

The delay operation just described pauses execution of an initial or always process for a fixed period of time. In a similar manner, Verilog also provides the ability to pause execution until a specified event occurs.

@(*event_expr*) *statement*;

An event occurs whenever the value of a net or register changes. In addition, all versions of Verilog support named events, which are explicitly created events that are discussed next. When executed, an ***event statement*** will block execution while it monitors one or more nets, registers, or named events. As soon as the desired event occurs, the event delay terminates and control is passed to the next statement. Changes that occur prior to when control passes to the event statement are ignored. Also ignored are old values being re-assigned to a net or register that is being monitored while the event statement waits for an event. Thus, an assignment to a net or register only creates an event if it causes a change in value.

Named Events. A named event allows the evaluation of code to be triggered by remote events without the need to be aware of the details of the event, such as how it was generated. These events can even traverse module boundaries. To produce a named event, it is necessary to first declare it, then it must be triggered using the '–>' operator, finally it is sensed. This is illustrated in the following example, which is a 'wireless' clock generator. In this example, the module *clockgen* produces a named event *tick* every 100 time units. This event is sensed in the module *timestamp*, causing it to print the time.

Example:

```
module top;
    clockgen clock();
    timestamp ts();
endmodule

module clockgen;
    event tick;
    always
        #100 -> tick;
endmodule

module timestamp;
    always @clock.tick
        $strobe("%g", $abstime);
endmodule
```

Event Expressions. The event statement monitors changes in the value of nets and registers using an event expression. There are two possible operations that can be performed on events when involving them in an expression. They can be filtered or combined. The keywords *posedge* or *negedge* applied to a net or register will filter out events associated with undesired changes. In particular, *posedge d* produces an event when *d* transitions from 0 to 1, from 0 to *x*, or from *x* to 1; *negedge d* produces an event when *d* transitions from 1 to 0, from 1 to *x*, or from *x* to 0. A d-flip-flop example shows how to use events to produce an ***edge triggered*** model.

```
module dff (q, d, clk);
    input clk, d;
    output q;
    reg q;

    always @(posedge clk)
        q = #5 d;
endmodule
```

Events are combined using the *or* keyword, meaning that an event on '*a or b*' occurs whenever there is an event on either *a* or *b*.

In Verilog-AMS event expressions found in initial and always processes may contain the analog event monitor functions described in Section 6.8 on page 204. In this way, events that occur in analog signals can affect the behavior of initial and always processes. Similarly, the event expression described in this section can be used in analog processes.

Event statements work differently between initial and always processes and analog processes (5§6.8p204). In initial and always processes, execution pauses at the event statement until the desired event occurs, at which point execution continues starting at the next statement. In an analog process, execution does not pause at the event statement. Rather, execution simply bypasses the next statement except at those instants when the desired event occurs.

7.6.3 Waits

Whereas the event statements are used to implement models that wait for edges, the *wait* statements are used to implement ***level triggered*** models.

```
wait (expr) statement;
```

The wait statement takes one argument. If the value of the argument is 1 the subsequent statement is immediately executed. If the value of the expression is 0, execution is blocked until the argument becomes 1.

Both the event and wait statements watch for a situation that is generated by an external process. The difference between the two is that the event statement is edge-triggered whereas the wait statement is level-triggered. So if *a* is 1, then

```
wait (a) b = 1;
```

will immediately set *b* to 1 whereas

```
@(posedge a) b = 1;
```

will wait until *a* transitions away from 1 and then back again before setting *b* to 1.

7.6.4 Delayed Assignment

If the statements for introducing various forms of delay that were presented earlier in this section are combined with an assignment statement, they will delay both the evaluation of the expression on the right-hand side of the assignment, as well as the updating of the target variable.

Examples:

```
#10 a = b;
@(posedge clk) q = d;
```

What is often required is that the delay be inserted between the evaluation of the right-hand side and the update of the target. Verilog provides various forms of delayed assignment for these cases. To provide a fixed delay in a procedural assignment, insert the delay specifier immediately after the assignment operator.

Examples:

```
a = #10 b;
a <= #10 b;
```

The first example is equivalent to the following code fragment

```
begin
    tmp = b;
    #10 a = tmp;
end
```

In the second example, the value of *b* is recorded, an update to *a* is scheduled for 10 time units in the future, and then execution passes to the next statement. Then, 10 time units later, the value of *a* is updated to the saved value of *b.* The behavior of both of these examples is the genesis for the names blocking and non-blocking assignment. In the first example, the process blocks waiting for the delayed assignment, in the second example the process does not block.

Similarly, one could delay the assignment until a particular event by inserting an event specifier immediately after the assignment operator.

Examples:

```
q = @(posedge clk) d;
a <= @(posedge c) b;
```

The first example is equivalent to the following code fragment

```
begin
    tmp = d;
    @(posedge clk)
        q = tmp;
end
```

Delay is applied to continuous assignments and net assignments by placing the delay specifier before the target.

Examples:

```
assign #5 a = ~b;
assign #5 a = ~b, c = ~d;
wire #10 c = a ^ b;
```

7.7 Conditionals

The conditionals available in initial and always processes are the same as those available in analog processes. They are described in Section 6.5 on page 200.

7.8 Iterators

The iterators available in initial and always processes are largely the same as those available in analog processes. They were described in Section 6.6 on page 202. The one iterator allowed in initial and always processes that is not allowed in analog processes is the forever loop.

7.8.1 Forever Loop

The forever loop simply repeats the statement that follows it forever.

forever
    statement;

An example of a module that uses a *forever* loop is given on page 212.

7.9 User-Defined Functions and Tasks

Verilog provides two ways of encapsulating code into reusable blocks within a module: the function and the task. Functions are provided to allow code to be shared, whereas tasks are provided to allow shared hardware to be modeled. Functions and tasks are described more fully next, with Table 14 highlighting the differences.

TABLE 14 *The distinguishing characteristics of tasks and functions.*

Tasks	Functions
Permits time control	Evaluates in a single simulation point (no state)
Local variables retain the values between invocations.	Local variables do not retain their values between invocations.
May have zero arguments	Must have a least one argument
Does not return a value, assigns values to output and inout parameters	Returns a single value. Output parameters are not allowed.
May contain #, @, –>, wait, task	Must not contain #, @, –>, wait, task

7.9.1 Functions

A function is code that is encapsulated and parameterized so that it can be shared throughout a module. As such, it is similar to functions in conventional programming languages. Functions take one or more arguments and return a single value. They must be defined within a module and are called from expressions. A function may declare its own variables, but unlike variables in modules and tasks, the values are not retained between calls to the function. A function cannot contain any timing control, (#, @ or *wait*); nor can they call tasks.

A function definition begins with the keyword *function*, optionally followed by the type of the return value of the function, then the name of the function and a semicolon. One or more input arguments may be declared along with any number of variables. The body of the function consists of a single statement that follows the declarations, and the function ends with the keyword *endfunction*. The default return type is a logic value. The return value is set by assigning a value to a variable whose name is the same as the name of the function.

Example:

```
function [15:0] mux;
    input [15:0] in1, in2, in3, in4;
    input [1:0] select;

    case (select)
        2'b00: mux = in1;
        2'b01: mux = in2;
        2'b10: mux = in3;
        2'b11: mux = in4;
    endcase
endfunction
```

7.9.2 Tasks

A ***task*** represents hardware that is encapsulated and parameterized so that it can be shared throughout a module. As it represents shared hardware, the values of any variables are retained at all times. Furthermore, each task represents a single piece of hardware, meaning that there is only one set of variables.

Tasks take zero or more parameters that can be input, output, or inout. When called, the values of input and inout parameters are copied into the task, and when the task terminates and returns, the values of the output and inout parameters are copied into the calling process.

Tasks must be defined within a module and they are called as an independent statement and not as part of an expression. Unlike functions, a task can contain timing control (#, @ or *wait*) and can call tasks. A task can call itself, or be called from other tasks that it has called. It can even be called while it is still running as a result of a previous call from another process.

Example:

```
module top();
    integer fib;

    initial repeat (30) begin
        #1 fibonacci(fib);
        $strobe("%d", fib);
    end

    task fibonacci;
        output out;
        integer out, prev, prev2;

        begin
            if (prev === 'hx) begin
                prev = 0;
                prev2 = 1;
                out = 0;
            end else begin
                out = prev + prev2;
                prev2 = prev;
                prev = out;
            end
        end
    endtask
endmodule
```

8 Mixed Behavior

Purely continuous time behavioral models are written using an analog process. Purely discrete-event behavioral models, either digital or analog, use initial or always processes. In addition, there are times when mixed behavioral models are needed. Either continuous time models driven by discrete-event processes or visa versa. Verilog-AMS supports this by allowing restricted access to continuous time nets and variables from a discrete-event process and to discrete-event nets and variables from a continuous-time process. There are four types of mixed-behavior access.

- Monitor the value of discrete-event nets and variables from an analog process.
- Monitor the value of continuous-time signals and variables from an initial or always process.

- Detect discrete events from within an analog process.
- Detect continuous-time events from within an initial or always process.

Before describing these access methods, it is important to again mention the issue of variable capture. Any variable or register that is assigned a value from within an analog process is captured by that process, meaning that it is not possible to assign it a value outside that process. Such variables are said to be captured or owned by the analog kernel. Conversely, if a variable is assigned a value outside an analog process, say in a initial or always process, it is not possible to also assign it a value from within an analog process. Such variables are said to be captured or owned by the discrete-event kernel.

8.1 Discrete-Event Values in an Analog Process

From within an analog process it is not possible to set the value of a discrete-event net, such as a wire or wreal, or a variable or register captured by the discrete-event kernel, but it is possible to read the value of such objects. Furthermore, if an analog process is actively monitoring the value of a discrete-event signal or variable, the process evaluates whenever the value of the signal or variable changes. Consider the following simple 16-bit digital-to-analog converter model.

```
module dac (out, in);
    parameter fullscale = 1.0;
    input [15:0] in;
    electrical out;

    analog
        V(out) <+ in * (fullscale/65536);
endmodule
```

In this module, the value of the16-bit bus *in* is cast to an integer before it is used. In this case it would be an error if *in* contained any bits that had values of *x* or *z*, and so with this model it is important to use it in a setting where that could not occur.

In this example, the analog process is sensitive to the value of *in* at all times, meaning that in addition to the times when it evaluates for other reasons, the process is guaranteed to evaluate immediately after a change in the value of *in*. This is not true in all cases, because in some cases the analog process is only sensitive to the value associated with the discrete-event kernel at isolated instants in time. Consider the following modification to the above example.

```
module dac (out, in, clk);
    parameter fullscale = 1.0;
    input [15:0] in;
    input clk;
    electrical out;
    real smpld;

    analog begin
        @(posedge clk)
            smpld = in * (fullscale/65536);
        V(out) <+ smpld;
    end
endmodule
```

In this case the analog process is only sensitive to the value of *in* at the instant of a rising edge on *clk*, and so the value on *in* can change without causing the process to evaluate.

Accessing a value of *x* (unknown) or *z* (high impedance) from a register or net in an analog process is an error except when accessed by specific constructs that allow it. These constructs can be used to convert the undesirable values to acceptable values before further processing. In particular, one can use the equality ('===') or inequality ('!==') operators to test values that may potentially contain an *x* or *z*, or one may use *case*, *casex*, or *casez* statements on values that may contain an *x* or *z*. One may also use binary, octal, or hexadecimal constants that may contain an *x* or *z* as arguments to these operators and statements. These ideas are used to create a analog tristate buffer.

```
module buf3 (out, in);
    input in;
    output out;
    electrical out;
    real value;

    analog begin
        @(in) case (in)
            1'b0: value = 0;
            1'b1: value = 1;
        endcase
        if (in !== 1'bz)
            V(out) <+ value;
    end
endmodule
```

This module evaluates the case statement whenever *in* changes and will set the local variable *value* to 0 if *in* is 0 and to 1 if *in* is 1. Otherwise it leaves *value* unchanged.

Finally, it tests the value of *in* and if not *z* it drives *V(out)* with *value*, otherwise it leaves *out* un-driven.

8.2 Discrete Events in an Analog Process

Notice that in the previous two examples, a discrete event was the target of an @ statement in an analog process. This is always possible. Any event, regardless of its source, may be used in any event expression, regardless of where it resides.

8.3 Continuous-Time Values in an Initial or Always Process

The value of continuous-time signals and variables can be read from within a discrete-event process without restriction. The following is an example where a continuous time signal *V(in)* is sampled with an always process to create a sample-and-hold.

```
module sah (out, in, clk);
    input in, clk;
    output out;
    electrical in;
    wreal out;
    real smpld;

    always @(posedge clk)
        out = V(in);

    assign out = smpld;
endmodule
```

It is important to remember that accessing an analog value from within a discrete process does not cause the two processes to synchronize. In other words, the analog process does not necessarily place a time point at the time when one of its values is sampled from a discrete process so as to accurately resolve the value. In this case if it is desired that the value used for *V(in)* be very accurate an analog process should be added to control the time point selection. For example, adding

```
analog @(posedge clk)
    ;
```

would result in the analog kernel placing a time point at the time when *V(in)* was sampled.

8.4 Continuous Events in an Initial or Always Process

As mentioned above in Section 8.2, analog events may be used in an event expression found in either an initial or always process. The following example uses this to model a comparator.

```
module comparator (out, in);
    input in;
    output out;
    electrical in;
    reg out;

    always @(cross(V(in)))
        out = V(in) > 0.0;
endmodule
```

8.5 Calling Functions

Analog functions can only be called from an analog process. Other user defined functions can only be called from outside an analog process.

9 Hierarchy

Hierarchical hardware descriptions are supported by allowing modules to be instantiated within other modules. Higher-level modules create instances of lower-level modules and communicate with them through input, output, and bidirectional ports. The behavior of instantiated modules may be customized using parameters. The parameter values used for the instantiated module can be set ether when instantiating the module or from anywhere within the hierarchy using a *defparam* statement (5§9.4.2p233).

To describe a hierarchy of modules, the user provides textual definitions of various modules. Each module definition stands alone; the definitions are not nested. Statements within the module definitions create instances of other modules, thus describing the hierarchy.

9.1 Modules

A module definition is enclosed between the keywords *module* and *endmodule*. The identifier following the keyword module is the name of the module being defined. It must be unique in that no other module or primitive can have the same name. The optional list of ports specify an ordered list of the module's ports. The order used can be significant when instantiating the module. The module may also declare a set of parameters using *parameter* statements. The order in which the parameters are declared may also be significant when instantiating a module.

The keyword *macromodule* can be used interchangeably with the keyword *module* to define a module. An implementation can choose to flatten the hierarchy of macromodules for efficiency.

A *top-level module* is a module that is defined but never instantiated by any other modules.

Behavioral descriptions would be found in the *analog*, *initial*, and *always* processes or in continuous assignments within the module. If the module is self-contained in that it does not instantiate any other modules, the module is considered a behavioral module. If the module instantiates other modules but does not contain *analog*, *initial*, or *always* processes it is considered a structural module. It is also possible for a module to contain both structural descriptions (instances of other modules) and behavioral descriptions (includes processes), or for it to contain neither. A module that contains neither generally contains constants that are accessed from other modules using hierarchical references (5§9.4p230).

9.2 Instantiation

Instantiation allows one module to incorporate a copy of another module into itself. Module definitions do not nest. That is, one module definition does not contain the text of another module definition within its *module-endmodule* keyword pair. A module definition nests another module by referencing or ***instantiating*** it.

Consider the module definition of the comparator given in Listing 5.

LISTING 5 *A comparator.*

```
module comparator (out, pin, nin);
    parameter real td = 1n from (0:inf);      // time delay (s)
    parameter real tt = 1n from [0:inf);      // output transition time (s)
    output out;
    input pin, nin;
    voltage out, pin, nin;
    real Vout;

    analog begin
        @(cross(V(pin) – V(nin), 0))
            ;
        Vout = ((V(pin) > V(nin)) ? 1 : 0);
        V(out) <+ transition(Vout, td, tt);
    end
endmodule
```

A single instance of this module might be created as follows,

```
    comparator C1(out, in, gnd);
```

In this case, an instance of the module *comparator* is created and named *C1*. The name is optional.

```
comparator (out, in, gnd);
```

The ports of the module are connected to nets *out*, *in* and *gnd*, with the association being done using the order in which the ports were defined. It is also possible to connect the nets to ports using the name of the ports,

```
comparator C1(.pin(in), .nin(gnd), .out(out));
```

In this example net *in* is connected to port *pin*, net *gnd* is connected to port *nin*, and net *out* is connected to port *out*. Notice that in this case the nets were given in a different order.

It is also possible to pass parameters into the module when instantiating it.

```
comparator #(1n, 2n) C1(out, in, gnd);
```

As with ports, the argument values can be associated with parameters either by order or by name. In the above example they are associated by order, and below they are associated by name.

```
comparator #(.tt(2n), .td(1n)) C1(out, in, gnd);
```

If a module is defined without ports, or if all the ports are to be left unconnected, the list of connections would be empty, but the parentheses are still required to be present.

Items can be left unspecified in both port and parameter lists. If passing by name, simply do not supply the name or the value. When passing by order, simply leave out the value. If the value is not the last in the list, the commas used to separate the values must still be provided even though the value itself is not (3§2p41). For example, with the comparator above, to give the value of *td* by order without giving *tt*, use

```
comparator #(, 2n) C1(out, in, gnd);
```

Instantiation occurs within a module definition, but outside any *analog*, *initial*, or *always* processes.

Example:

```
module smpl_ckt;
    electrical n;
    ground gnd;

    vsrc #(.dc(1)) V1(n, gnd);
    resistor #(.r(1k)) R1(n, gnd);
endmodule
```

9.2.1 Multiple Instantiation

A single instance statement may instantiate more than one instance of a module. There are two ways in which this might happen. In the first, one simply gives additional instance names (optional) and port lists on the instance line. For example, one could terminate a bus with the following collection of resistors,

```
resistor #(50) r1 (bus[1], gnd),
               r2 (bus[2], gnd),
               r3 (bus[3], gnd),
               r4 (bus[4], gnd),
               r5 (bus[5], gnd),
               r6 (bus[6], gnd),
               r7 (bus[7], gnd),
               r8 (bus[8], gnd);
```

With the very regular nature of this case it is possible to express this same set of resistors even more succinctly with the other form of multiple instantiation, arrays of instances.

```
resistor #(50) r[1:8] (bus, gnd),
```

In this example, an array of eight instances is created, *r[1]* to *r[8]*. One each is connected to the members of *bus* and all are connected to *gnd*. Other than being integers, there are no constraints on the array bound. In particular, one bound need not be larger than the other, and indeed they may both be the same, in which case only one instance is generated. If a scalar net is passed to a port of an array of instances, then it is connected to that port on each of the instances. If the net being passed in is an array, then the members of the net are distributed to the instances in order. If the port is a scalar, then it will consume one member of the net. If it is an array itself, then it will consume multiple members of the net. In the end, all members of both the net and the ports must be consumed.

One can create multidimensional arrays of instances by creating arrays of arrays of instances.

9.3 Gate-Level Descriptions

Verilog provides a set of predefined ***gate level primitives*** as shown in Table 15. These primitives can be instantiated just like modules. The latch below illustrates the use of these gate level primitives.

TABLE 15 *Gate and switch level primitives*

multiple input gates	multiple output gates	tristate gates	pull gates	MOS switches	bidirectional switches
and	buf	bufif0	pullup	nmos	tran
nand	not	bufif1	pulldown	pmos	tranif0
nor		notif0		cmos	tranif1
or		notif1		rnmos	rtran
xor				rpmos	rtranif0
xnor				rcmos	rtranif1

```
module latch (q, qb, rb, sb);
    input sb, rb;
    output q, qb;

    nand    (q, sb, qb),
            (qb, q, rb);
endmodule
```

sb
q
rb
qb

This example instantiates two unnamed nand gates. By convention, output ports are specified before input ports on built-in primitives. Delay is added to the gates by adding a delay specifier after the gate's module name.

```
module latch (q, qb, rb, sb);
    input sb, rb;
    output q, qb;

    nand #5 (q, sb, qb),
            (qb, q, rb);
endmodule
```

9.4 Hierarchical Names

9.4.1 Rules of Scope

The names used in a context need not be declared within that context. When a name is encountered the local context is searched for a declaration. If the name is not found, the search is continued by moving higher in the instantiation hierarchy (closer to the root or top-level module). The hierarchy itself is made up of modules and named blocks. So if a name is encountered in a named block, that block is searched for declarations for that name. If not found, the context that contains the named block is searched. That may either be a module definition or another named block. If still not found, the search continues until it eventually leaves the module definition. At that point it progresses up the instantiation hierarchy. So the module that instantiates the module with the name is searched, and so on. Eventually, the search may reach the top-level modules where it ends. If still not found, the name is undeclared, which is an error. Once the search leaves the module where the name was originally used, it will not find particular types of objects, such as nets, variables, and parameters. These objects must be declared local to the module.

```
module alpha;
    integer a;
    gamma();

    analog function real uno;
        input arg;
        real arg;
        begin
            uno = arg;
        end
    endfunction
endmodule

module beta;
    integer b;

    analog function real dos;
        input arg;
        real arg;
        begin
            dos = arg;
        end
    endfunction
endmodule

module gamma;
    integer c, d;
    analog begin
        begin : delta
            integer d;
            d = 1;          // this d is the one declared in the named block delta
            c = 1;          // this c is the one declared in the module gamma
            d = uno(1);     // this uno is the function defined in alpha, it is
                            // available because gamma is instantiated from alpha
            a = 1;          // this is an error, the a in alpha is unavailable because
                            // it is a variable outside of module scope
```

```
            d = dos(1);  // this is an error, dos is defined in beta, which is not in
                         // the instantiation path for gamma and is not available
        end
    end
endmodule
```

Hierarchical Names

Hierarchical names consist of a path name of identifiers separated by periods ('.'). The first identifier is found by searching up the procedural and module hierarchy. From where the first identifier is found, each succeeding identifier in the path name specifies the named scope with which to continue searching downward. The last identifier specifies the entity being searched for.

With hierarchical names and the rules of scoping, objects often have multiple names, and those names depend on where the name is found. For example, objects in the example have the following names from within the named block *delta* if module *alpha* is instantiated as *alpha1* and *beta* is instantiated as *beta1*,

Objects defined in module *alpha*,

alpha1
alpha1.a
uno
alpha1.uno

Objects defined in module *beta*,

beta1
beta1.b
beta1.dos

Objects defined in module *gamma*,

d (this is the *d* in *delta*)
c
gamma
gamma.c
gamma.d (this is the *d* defined in *gamma* outside *delta*)
delta.d (this is the *d* in *delta*)
gamma.delta.d (this is the *d* in *delta*)
alpha.gamma
alpha.gamma.d (this is the *d* in *delta*)
alpha.gamma.c
alpha.gamma
alpha.gamma.d (this is the *d* defined in *gamma* outside *delta*)
alpha.delta.d (this is the *d* in *delta*)

alpha.gamma.delta.d (this is the *d* in *delta*)

From a stylistic perspective it is generally preferred to use the shortest of the various names that can be used for a particular object.

A hierarchical name that refers to an object outside of the module where the name is found is often referred to as an OOMR (rhymes with goo-mar), which is short for out-of-module-reference.

From within an analog process block, it is possible to use hierarchical name referencing to access signals on an external branch, but not external analog variables or parameters. When accessing external branches, a branch signal (its potential or flow) can be monitored (probed); for source branches, contributions can be made to the output signal.

9.4.2 Defparams

Using the *defparam statement*, parameter values on any module instance throughout the design can be set using the hierarchical name of the parameter.

Consider a situation where an instance of *receiver RX1* contained two instances of *filter*, *Ifilter* and Qfilter, and that *filter* supported a parameter *bw*. Those parameters could be set from a top-level module as follows.

```
module setparams;
    defparam    RX1.Ifilter.bw = 2.5M,
                RX1.Qfilter.bw = 2.5M;
endmodule
```

The expression on the right-hand side of *defparam* assignments must be a constant expression involving only constant numbers and references to parameters. If the expression contains parameters, they must be declared in the same module that contains the *defparam* statement.

The *defparam* statement is particularly useful for grouping all of the parameter value override assignments together in one module.

9.5 Mixed Signal Structure

Mixed-signal structure is presented in detail in Section 4 in Chapter 4 (beginning on page 121) and so will not be discussed further here.

10 Other Features of Verilog-HDL

Many features of the Verilog language are beyond the scope of this book. In particular, the following topics are not covered in this book, but you should be able to find out about them by reading any of the available Verilog-HDL books [1, 5, 23, 27] or the Verilog-HDL language reference manual [16].

- strengths and switch-level modeling
- complex delay specifications
- user defined primitives (UDPs)
- integer time variables
- numerous compiler directives
- numerous system tasks and functions
- specify blocks, specparams
- pin to pin path delays and state dependent path delays (*specify*)
- timing constraint checks (ex. *setupandhold*)
- Verilog Procedural Interface (VPI) routines

What's Next

This chapter completes the presentation of Verilog-A/MS. The only thing that remains is to discuss the use of some of the implementations of Verilog-A/MS. The appendix presents many of the practical details that must be understood when using simulators that support Verilog-A/MS.

A Compatibility

This appendix presents some of the practical details that you need to know to be able to use Verilog-A/MS with the simulators available that support it. It starts by describing what constructs to avoid with purely digital models in order to retain compatibility with traditional Verilog-HDL simulators. Then, the issues involved when using SPICE models from Verilog-A/MS are discussed. Finally, it describes specifically how to use Verilog-A and Verilog-AMS in the Cadence simulators: *Spectre* and *AMS Designer.* Spectre was the first simulator to support Verilog-A. It is currently the most popular simulator, and was used to validate each of the Verilog-A models contained in this book. Similarly, AMS Designer was the first simulator to support Verilog-AMS, is currently the most popular simulator that does so, and was used to validate each of the Verilog-AMS models given in this book.

1 Verilog-HDL Compatibility

When describing purely digital modules, it is often desirable to completely avoid the use of the AMS extensions so that the models can be read by a Verilog-HDL simulator without modification. In these cases, one should consciously avoid the following constructs.

1. Explicit parameter type declarations
2. Parameter range limits
3. Analog declarations such as disciplines, natures, branches and ground.
4. Analog processes and the various things associated with them (contributions, analog operators, limiting and stimulus functions, etc.)
5. Analog events, such as *cross*, *timer*, *initial_step* and *final_step*.
6. Analog functions
7. System functions and tasks that were added to support analog or mixed-signal modeling, such as $*abstime*, *analysis*, $*bound_step*, $*discontinuity*, the $*driver_*... functions, $*limexp*, the $*rdist_*... functions, $*temperature*, and $*vt*.
8. The *wreal* wire type

9. Connect modules and connect rules

2 SPICE Compatibility

At some point all circuit simulators such as SPICE will understand Verilog-A, all device models will be available as Verilog-A modules, and model files and netlists will be readily available in Verilog formats; but not today. Until that day, simulators that support Verilog-A/MS must provide the ability to access SPICE primitives from a Verilog description. This section describes how access to SPICE built-in primitives is provided from the Verilog-A/MS language.

2.1 Scope of Compatibility

SPICE is not a single language, but rather is a family of related languages. The first widely used version of SPICE was SPICE2g6 from the University of California at Berkeley. However, SPICE has been enhanced and distributed by many different companies, each of which has added their own extensions to the language and the models. As a result, there is a great deal of incompatibility even among the SPICE languages themselves.

Verilog-A/MS makes no judgment as to which of the various SPICE languages should be supported. Instead, it states that if a simulator that supports Verilog-A/MS is also able to read SPICE netlists of a particular flavor, then certain objects defined in that flavor of SPICE netlist can be referenced from within a Verilog-A/MS structural description. In particular, SPICE models and subcircuits can be instantiated within a Verilog-A/MS module. This is also true for most of the SPICE primitives that are built into the simulator.

2.1.1 Degree of Incompatibility

There are four primary areas of incompatibility between versions of SPICE simulators.

1. The version of the SPICE language accepted by various simulators is different and to some degree proprietary. This issue is not addressed by Verilog-A/MS. So whether a particular Verilog-A/MS simulator is SPICE compatible, and with which particular variant of SPICE it is compatible, is solely determined by the authors of the simulator.
2. Not all SPICE simulators support the same set of component primitives. A particular SPICE netlist may reference a primitive that is unsupported within a given simulator. Verilog-A/MS offers no alternative in this case other than the possibility that if the model equations are known, the primitive can be rewritten as a module.

3. The names of the built-in SPICE primitives, their parameters, or their ports can differ from simulator to simulator. This is particularly true because many primitives, parameters, and ports are unnamed in SPICE. When instantiating SPICE primitives in Verilog-A/MS, the primitives must, and parameters and ports can, be named. Since there are no established standard names, there is a high likelihood of incompatibility cropping up in these names.

   To avoid this, Verilog-A/MS defines a list of names that must be supported for common SPICE primitives when made available within Verilog-A/MS. This list is given in Table 1. However, it is not possible to anticipate all SPICE primitives and parameters that could be supported; so different implementations can end up using different names. This level of incompatibility can be overcome by using wrapper modules to map names.
4. The mathematical description of the built-in primitives can differ. As with the netlist syntax, incompatible enhancements of the models have crept in through the years. Again, Verilog-A/MS offers no solution in this case other than the possibility that if the model equations are known, the primitive can be rewritten as a module.

2.2 Accessing SPICE Objects from Verilog-A/MS

If an implementation of a Verilog-A/MS tool supports SPICE compatibility, it is expected to provide the basic set of SPICE primitives given in Section 2.3 and be able to read SPICE netlists that contain models and subcircuit statements.

SPICE primitives built into the simulator are treated in the same manner in Verilog-A/MS as built-in primitives. However, while the Verilog-A/MS built-in primitives are standardized, the SPICE primitives are not. All aspects of SPICE primitives are implementation dependent.

In addition to SPICE primitives, it is also possible to access subcircuits and models defined within SPICE netlists. The subcircuits and models contained within the SPICE netlist are treated as module definitions.

2.2.1 Case Sensitivity

SPICE netlists are case insensitive, whereas Verilog-A/MS descriptions are case sensitive. From within Verilog-A/MS, a mixed-case name matches the same name with an identical case as if it were defined in a Verilog description. However, if no exact match is found, the mixed-case name will match the same name defined within SPICE regardless of the case.

2.2.2 Examples

Accessing SPICE models. Consider the following SPICE model file being read by a Verilog-A/MS simulator.

```
.model vertnpn npn   bf=80 is=1e-18 rb=100 vaf=50
+                    cje=3pf cjc=2pf cjs=2pf tf=0.3ns tr=6ns
```

This model can be instantiated in a Verilog-A/MS module as follows

```
module diffPair (c1, b1, e, b2, c2);
    electrical c1, b1, e, b2, c2;

    vertNPN Q1 (c1, b1, e, );
    vertNPN Q2 (.c(c2), .b(b2), .e(e));
endmodule
```

Unlike with SPICE, the first letter of the instance name, in this case *Q1* and *Q2*, is not constrained by the primitive type. For example, they can just as easily be *T1* and *T2*. The ports and parameters of the BJT are determined by the BJT primitive itself and not by the model statement for the BJT. See Table 1 for more details. The BJT has 3 mandatory ports (collector, base, and emitter) and one optional port (the substrate). In the instantiation of *Q1*, the ports are passed by order. With *Q2*, the ports are passed by name. In both cases, the optional substrate port is defaulted by simply not giving it.

Accessing SPICE Subcircuits. As an example of how a SPICE subcircuit is referenced from Verilog-A/MS, consider the following SPICE subcircuit definition of an oscillator.

```
.subckt ecposc (out gnd)
    va vcc gnd 5
    iee e gnd 1ma
    q1 vcc b1 e vcc vertnpn
    q2 out b2 e out vertnpn
    l1 vcc out 1uh
    c1 vcc out 1p ic=1
    c2 out b1 272.7pf
    c3 b1 gnd 3nf
    r1 b1 gnd 10k
    c4 b2 gnd 3nf
    r2 b2 gnd 10k
.ends ecposc
```

This oscillator can be referenced from Verilog-A/MS as:

```
module osc (out, gnd);
    electrical out, gnd;
```

```
        ecpOsc Osc1 (out, gnd);
    endmodule
```

Note that in Verilog-A/MS the name of the subcircuit instance is not constrained to start with *X* as it is in SPICE.

Accessing SPICE Primitives. To show how various SPICE primitives can be accessed from Verilog-A/MS, the subcircuit given above is translated to native Verilog-A/MS.

```
    module ecpOsc (out, gnd);
        electrical out, gnd;

        vsine #(.dc(5)) Vcc (vcc, gnd);
        isine #(.dc(1m)) Iee (e, gnd);
        vertnpn Q1 (vcc, b1, e, vcc);
        vertnpn Q2 (out, b2, e, out);
        inductor #(.l(1u)) L1 (vcc, out);
        capacitor #(.c(1p), .ic(1)) C1 (vcc, out);
        capacitor #(.c(272.7p)) C2 (out, b1);
        capacitor #(.c(3n)) C3 (b1, gnd);
        resistor #(.r(10k)) R1 (b1, gnd);
        capacitor #(.c(3n)) C4 (b2, gnd);
        resistor #(.r(10k)) R2 (b2, gnd);
    endmodule
```

2.3 Preferred Primitive, Parameter and Port Names

Table 1 shows the required names for primitives, parameters, and ports that are otherwise unnamed in SPICE. For connection by order instead of by name, the ports and parameters are given in the order listed. The default discipline of the ports for these primitives is *electrical* and their direction is *inout*.

TABLE 1 *Required names for SPICE primitives.*[a]

Primitive Name	Port Names	Parameter Names
resistor	p, n	r, tc1, tc2
capacitor	p, n	c, ic
inductor	p, n	l, ic
vexp	p, n	dc, mag, phase, val0, val1, td0, tau0, td1, tau1

TABLE 1 *Required names for SPICE primitives.*[a]

Primitive Name	Port Names	Parameter Names
vpulse	p, n	dc, mag, phase, val0, val1, td, rise, fall, width, period
vpwl	p, n	dc, mag, phase, wave
vsine	p, n	dc, mag, phase, offset, ampl, freq, td, damp, sinephase, ammodindex, ammodfreq, ammodphase, fmmodindex, fmmodfreq
iexp	p, n	dc, mag, phase, val0, val1, td0, tau0, td1, tau1
ipulse	p, n	dc, mag, phase, val0, val1, td, rise, fall, width, period
ipwl	p, n	dc, mag, phase, wave
isine	p, n	dc, mag, phase, offset, ampl, freq, td, damp, sinephase, ammodindex, ammodfreq, ammodphase, fmmodindex, fmmodfreq
diode	a, c	area
bjt	c, b, e, s	area
mosfet	d, g, s, b	w, l, ad, as, pd, ps, nrd, nrs
jfet	d, g, s	area
mesfet	d, g, s	area
vcvs	p, n, ps, ns	gain
vccs	sink, src, ps, ns	gm
tline	t1, b1, t2, b2	z0, td, f, nl

a The names *diode*, *bjt*, *jfet*, *mesfet*, and *mosfet* are never used from within Verilog-A/MS because these components require a model. Thus, the model name is used in Verilog-A/MS instead of the primitive name.

2.3.1 Unsupported Components

Verilog-A/MS does not support the concept of passing an instance name as a parameter. As such, the following components are not supported: *ccvs*, *cccs*, and mutual inductors; however, these primitives can be instantiated inside a SPICE subcircuit that itself is instantiated in Verilog-A/MS.

2.4 Other Issues

There are currently some important unresolved issues with SPICE compatibility in Verilog-A/MS. It is expected that these issues will be resolved when extensions currently under development to support compact semiconductor models are added to the language.

2.4.1 SPICE Model Statements

There is currently no Verilog-A/MS equivalent to the SPICE model statement. The way a simulator accesses a library of SPICE models is implementation specific.

2.4.2 Multiplicity Factor on Subcircuits

SPICE simulators support a multiplicity factor (m) parameter on subcircuits without the parameter being explicitly declared. This factor is typically used to indicate the subcircuit should be modeled as if there are a specified number of copies wired in parallel. If supported by the implementation, the automatic multiplicity factors are supported for subcircuits defined in SPICE, but not for subcircuits defined as modules in Verilog-A/MS. Thus, if the SPICE subcircuit given in Section 2.2 is instantiated, a multiplicity factor could be specified (assuming the simulator implementation supports multiplicity factors on SPICE subcircuits). However, a multiplicity factor cannot be specified when instantiating the equivalent Verilog-A/MS module, also given in Section 2.2.

3 Spectre Compatibility

Spectre was the first simulator to support Verilog-A, and is still by far the most popular. It was also the simulator used to validate all of the Verilog-A models in this book.

3.1 Using Verilog-A with Spectre

When Spectre starts up, it reads either a Spectre or SPICE netlist. That netlist includes the top-level of the design, along with the various control statements. Those control statements specify the simulator settings and any analyses to be performed. In addition, the desired model and Verilog-A files would be referenced from this file.

When Spectre reads a reference to a Verilog-A file, it makes the components defined within that file available to the design. In other words, referencing a Verilog-A file does not instantiate the components described in the file into the design. Rather, it loads the components and makes them available for instantiation into the design. Once a Verilog-A component has been loaded into Spectre, it can be instantiated in the same manner as any other component. This is illustrated in the Spectre netlist shown below.

Example:

```
// Test circuit for quantizer
simulator lang=spectre
ahdl_include "quantizer.va"

Vclk (clk 0) vsource type=pulse val1=1 period=1us
Vin (in 0) vsource type=sine ampl=1 freq=5kHz sinephase=45
Quantizer (out in clk) quantizer levels=21 vh=1 vl=-1 vth=0.5 tt=10ns

save in out Quantizer:level
sineResp tran stop=200uS
```

The first line in this file is a comment, as it must be, and the second specifies that the Spectre netlist format is used in the file. The third line opens the Verilog-A file *quantizer.va* and loads the model found in this file. It contains the module *quantizer*, shown in Listing 1.

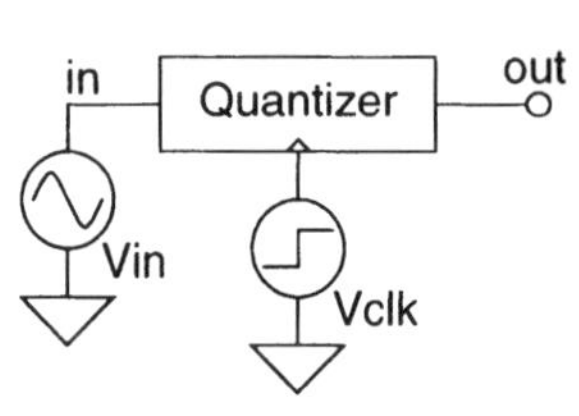

The next three lines instantiate the three components that make up the design. *Vin*, *Vclk*, and *Quantizer*. *Vin* and *Vclk* are Spectre primitive voltage sources and *Quantizer* is the Verilog-A model from Listing 1. Within Spectre and SPICE netlists, nodes are associated with ports by the order in which they are given. Thus for the quantizer, the nodes *out*, *in*, and *clk* are associated with the ports with the same names, but the fact that the names match is a coincidence, it is the order that determines the association. So the first node given is associated with the first port, and so on. Parameters are associated by name, and any parameters that are left unspecified take their default values.

The final two lines are control statements for Spectre. The first specifies which signals should be saved for later display or analysis. In this case, three signals are saved, *in*, *out* and *Quantizer:level*. The signals *in* and *out* are nodes, and Spectre will save the voltages on these nodes for each analysis. *Quantizer:level* saves the value of the variable *level* from *Quantizer* for each analysis. This is a feature of Spectre that can be used when debugging a module. The final statement specifies that a 200μs transient analysis named *sineResp* should be run.

If this netlist is contained in a file named *quantizer.scs* (the .scs suffix stands for *Spectre circuit simulator)*, it can be run in Spectre using

LISTING 1 *N-level quantizer model (like an ADC followed by a DAC).*

```
`include "disciplines.vams"

module quantizer (out, in, clk);
    parameter integer levels=2 from [2:inf);// number of quantization levels
    parameter real vh = +1;                 // voltage of highest level (V)
    parameter real vl = -1 from (-inf:vh);  // voltage of lowest level (V)
    parameter real vth = (vh + vl)/2;       // threshold voltage of clock (V)
    parameter integer dir = +1 from [-1:+1] exclude 0;
                                            // if dir=+1, rising clock edge triggers
                                            // if dir=-1, falling clock edge triggers
    parameter real td = 0 from [0:inf);     // output delay (s)
    parameter real tt = 0 from [0:inf);     // output transition time (s)
    output out; voltage out;    // output
    input in; voltage in;       // input
    input clk; voltage clk;     // clock input (edge triggered)
    real quantized, delta;
    integer level;

    analog begin
        @(cross(V(clk) - vth, dir) or initial_step) begin
            delta = (vh - vl)/(levels - 1);
            level = (V(in) - vl)/delta;
            if (level < 0)
                level = 0;
            else if (level >= levels)
                level = levels - 1;
            quantized = level * delta + vl;
        end
        V(out) <+ transition( quantized, td, tt );
    end
endmodule
```

```
    spectre quantizer.scs
```

from a Unix shell. More information about using Spectre can be found in [19].

Busses. Verilog-A allows busses as ports but the Spectre netlist language does not support busses. When instantiating a Verilog-A model that includes a bus as a port from within a Spectre netlist, each member of the bus must be individually listed in the terminal list. As an example, consider a netlist that instantiates the ADC and DAC given in Listings 26 and 27 from Chapter 3.

Example:

```
// Test ADC and DAC models
simulator lang=spectre

ahdl_include "adc2.vams"
ahdl_include "dac2.vams"

Vin (in 0) vsource type=sine ampl=0.5 dc=0.5 freq=1
Vclk (clk 0) vsource type=pulse val0=0 val1=1 period=1ms
ADC (b7 b6 b5 b4 b3 b2 b1 b0 in clk) adc bits=8 vdd=1 td=100us tt=100us
DAC (out b7 b6 b5 b4 b3 b2 b1 b0 clk) dac bits=8 vdd=1 td=100us tt=100us

sineResp tran stop=1
```

Here the 8-bit bus that connects the output of the ADC to the input of the DAC consists of 8 nodes, *b0* to *b7*, that are individually specified on the terminal list for both the ADC and DAC.

3.2 Accessing Spectre Objects from Verilog-A

In the last section, it was shown how to access Verilog-A models from a Spectre netlist. The converse is also possible; one can access Spectre primitives, subcircuits, and models from a Verilog-A netlist. This process was sketched out for SPICE in Section 2. Some of the details are a bit different with Spectre as it provides access to a broader variety of components and because the way Spectre implements its independent sources is a bit different from SPICE.

Any Spectre objects that are to be instantiated in a Verilog-A netlist must be defined, either in the Spectre netlist, or in the Spectre simulator itself, before they can be instantiated in a Verilog-A module.

There is no way to specify a model statement or model parameters (as opposed to instance parameters) from within the Verilog-A language. However, it is possible to instantiate an instance that references a model from within Verilog-A. For example the Spectre model

```
model vertNPN bjt type=npn bf=80 is=1e-18 rb=100 vaf=50 \
                    cje=3pf cjc=2pf cjs=2pf tf=0.3ns tr=6ns
```

can be instantiated in a Verilog-A/MS module as follows

```
module diffPair (c1, b1, e, b2, c2);
    electrical c1, b1, e, b2, c2;
    parameter real area=1 from (0:inf);
```

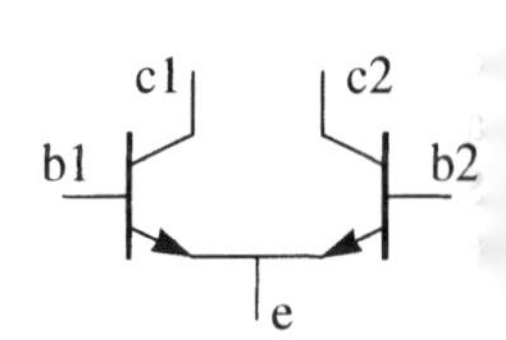

```
        vertNPN Q1 (c1, b1, e, );
        vertNPN Q2 (.c(c2), .b(b2), .e(e));
    endmodule
```

One thing to take notice of is that by default the Spectre netlist language is case sensitive, thus when referring to a name defined in the Spectre netlist from a Verilog netlist, the case of the characters used in the name must match between the two languages.

In a very similar manner, one can access subcircuits defined in a Spectre netlist from Verilog. Consider the above module implemented as a Spectre subcircuit.

```
    subckt diffPair (c1, b1, e, b2, c2)
        parameter area=1

        Q1 (c1 b1 e) vertNPN area=area;
        Q2 (c2 b2 e) vertNPN area=area;
    ends diffPair
```

This subcircuit can be instantiated in a Verilog-A netlist using

```
    diffPair #(.area(10)) DP1(o1, i1, tail, i2, o2);
```

Notice that you can specify values for Spectre subcircuit parameters.

Finally, any Spectre built-in components that use only real-valued, string-valued, or vector-valued parameters can be accessed from Verilog-A. This precludes direct use of current controlled and mutual inductors, but they can be used if bundled in subcircuits. The names used for the ports (terminals) and parameters can be determined by using Spectre's help facility. For example, to find the names needed to instantiate Spectre's multi-conductor lossy transmission line, run

```
    spectre -help mtline
```

at the Unix command prompt. It will describe the *mtline* component, and the description will give the names of the ports and parameters. From Verilog-A, the *mtline* can be instantiated with

```
    mtline #(.len(0.01), .r({0.3, 0.0, 0.3}), .c({0.35p,  0.03p, 0.35p})) x1 (a1, b1, a2, b2, gnd, gnd);
```

Spectre provides access to its independent sources in a manner somewhat different from that described in Section 2.3. Rather than encoding the wave shape in the name of the primitive, Spectre bases the name of the primitive only on the type of source, and adds an additional parameter that specifies the wave shape. For example, Section 2.3 indicates that instantiating a sinusoidal voltage source from SPICE would be done as follows,

```
    vsine #(.ampl(1), .freq(1G)) Vin (in, gnd);
```

but with Spectre you would use

```
vsource #(.type("sine"), .ampl(1), .freq(1G)) Vin (in, gnd);
```

The benefit of this approach is that the wave shape itself is parameterized. It is possible to give the parameters for multiple wave shapes, and then select which wave shape to use with the *type* parameter. The list of Spectre source parameters is given in Table 2.

TABLE 2 *Spectre independent source parameters.*

Mode	Name	Description
all modes	type	Waveform type, possible values are "dc", "pulse", "pwl", "sine", or "exp".
	mag, phase	Small signal level and phase.
"dc"	dc	DC level.
"pulse"	val0, val1	Pulse levels.
	delay	Start time of first pulse.
	rise, fall	Pulse rise and fall time.
	width	Pulse width.
	period	Pulse period.
"pwl"	wave	Vector of time/value pairs that define the waveform.
	file	A string that gives the name of a file that contains a list of time-value pairs that defines the waveform.
"sine"	dc	DC level of sinusoid.
	ampl	Amplitude of sinusoid.
	freq	Frequency of sinusoid.
	delay	Start time of first pulse.
	damp	Damping factor of sinusoid.
	sinephase	Phase of sinusoid.
	ammodindex	AM index of modulation.
	ammodfreq	AM modulation frequency.
	ammodphase	AM modulation phase.
	fmmodindex	FM index of modulation.

TABLE 2 *Spectre independent source parameters.*

Mode	Name	Description
	fmmodfreq	FM modulation frequency.
"exp"	val0, val1	Equilibrium levels.
	td0, td1	Start time for transitions to *val0*, *val1*.
	tau0, tau1	Time constant for transition to *val0*, *val1*.

3.3 Spectre's Implementation of Verilog-A

Currently Spectre supports a somewhat restricted version of the Verilog-A language. In particular, the following limitations exist in the Spectre implementation.

1. When instantiating modules, parameter values can only be passed by name and not by order. Furthermore, nets can only be associated with ports by order, not by name.
2. The *genvar* type of integers is not supported. Nor are analog operators allowed in *for* loops. Instead the deprecated *generate* statement remains available and should be used when analog operators must be contained within a loop.
3. The *ground* statement is not supported. You should pass in ground as a terminal.
4. Argument and port lists cannot contain skipped arguments.
5. Using parameter values as array sizes is not allowed.
6. It is not possible to pass strings as parameters into a Verilog-A module.
7. When declaring vector ports, array bounds must be given on both type and direction declarations.
8. The *delay* and *zi_...* functions do not work in any small-signal frequency-domain analyses.
9. The small signal stimulus functions cannot be used in assignment statements or in expressions. They must instead be isolated in a contribution statement. So, rather than scaling the output of the noise functions, you should specify the desired output power using the primary operand of the functions. For example, consider modeling oscillator phase noise for use with SpectreRF. Assume that *ampl*, *freq*, and *pn* are parameters and that *phase* is a real variable set such that

   ```
   phase = 2*`M_PI*freq*$abstime;
   ```

   Then the simplest way of modeling phase noise is to simply add noise to the argument used to convert phase to voltage, in this case cos().

```
V(out) <+ ampl*cos(phase + flicker_noise(pn,2));
```

But this is not supported as the *flicker_noise* function is not isolated on the contribution statement. Since the *flicker_noise* output is by definition small, it can be removed from the argument of the cos() function by using the Taylor series to expand the function into a power series and then truncating the higher order terms. This naturally leads to something like

```
V(out) <+ ampl*(cos(phase) + sin(phase)*flicker_noise(pn,2));
```

However, even this is not supported as the output of the *flicker_noise* function is being further processed before it reaches the contribution operator. Instead, both factors that are applied to the *flicker_noise* function must be applied directly to the operand.

```
V(out) <+ ampl*cos(phase) + flicker_noise(pn*ampl*sin(phase),2);
```

This is acceptable even though the *flicker_noise* function is not alone in the contribution statement because it is isolated.

10. The output of the derivative operator, *ddt*, must have linear accessibility to a branch. Meaning that its output can be directly contributed to a branch, or it can be scaled by a constant factor, but it cannot be multiplied by a signal dependent quantity or passed through a nonlinear operator before being contributed to a branch. So for example,

```
I(p,n) <+ ddt(c*V(p,n));
```

```
I(p,n) <+ c*ddt(V(p,n));
```

```
icap = c*ddt(V(p,n));
I(p,n) <+ V(p,n)/r + icap;
```

are all acceptable, but

```
I(p,n) <+ c0*(1+c1*(V(p,n) – v0))*ddt(V(p,n));
```

is not because the output of the *ddt* operator is being multiplied by a quantity that is dependent on the value of a continuous-time signal. To avoid this problem the model must either be reformulated [20] or a node must be created internal to the model to hold the output of the *ddt* operator. Then the nonlinear operation would be performed on the value of the node rather than directly on the output of the *ddt* operator.

4 AMS Designer Compatibility

Cadence's AMS Designer is both a simulator and an environment. The simulator implements the full Verilog-AMS language and was used to validate the mixed-signal models provided in this book. The environment is the interface between the Cadence Design Framework II (DFII) and the simulator. It is used to netlist schematics into Verilog-AMS format, to create the design configurations, and to run the simulator. The Verilog-AMS language is the primary design language within AMS designer; meaning that it is used for both netlisting and as the modeling language. In addition, AMS Designer is a multilingual simulator that also supports VHDL-AMS, SystemC, SystemVerilog, and Spectre and SPICE netlists.

The following describes how the examples in this book can be used with the simulator by providing necessary instruction on how to set-up an AMS simulation. Basic instructions are given for using the AMS Designer simulator in a stand-alone mode. The product documentation available from Cadence provides complete and detailed help on how to use the simulator as well as instruction on how to use the AMS Designer environment.

4.1 Using Verilog-AMS with AMS Designer

The AMS Designer simulator is built upon NC-SIM technology. Both share a common library concept and format.

4.1.1 General Setup

To use a Verilog-AMS model it must be compiled into one of the libraries defined in the *cds.lib* file (see Listing 2) located in the working directory. In this file, '#' is used to introduce a comment: anything that follows up until the end of the line is ignored. The file consists of commands, one per line. The *include* command is replaced by the contents of the file it references. In this case, it includes the contents of the *cds.lib* file that comes with the simulator installation. This file references the libraries that are delivered with the simulator. An example of such a library is *connectLib*, which contains definitions of various interface components with different accuracy levels. The '$' introduces a Unix environment variable substitution. In this case, *AMSHOME* would contain the full path to the installation directory for the AMS Designer simulator. The *define* command declares that the Unix directory *worklib* in the working directory shall be used as the library directory for the library with the name *worklib*.

LISTING 2 *Sample cds.lib file*

```
# cds.lib
include $AMSHOME/tools/inca/files/cds.lib
define worklib ./worklib
```

Another important setup file is *hdl.var.* Place tool options that are often used into this file rather than specifying them on the command line every time a model file is compiled or a design is elaborated. Listing 3 shows an example *hdl.var* file. The *include* statement incorporates the default simulator settings. The first *define* statement declares that the name *worklib*, which was declared within the *cds.lib* file, is defined as the working library. *WORK* is a keyword, *worklib* is the name of the existing library. This causes models to be compiled into this library if no other target library is specified as a command line argument during compilation. The second define statement causes the compiler to write information into the library necessary for source code debugging.

LISTING 3 *Sample hdl.var file*

```
# hdl.var
include $AMSHOME/tools/inca/files/hdl.var
define WORK worklib
define NCVLOGOPTS -linedebug
```

4.1.2 Compilation

Assume that the *cds.lib* and *hdl.var* files given above are found in a directory that contains the Verilog-AMS model *vco.vams*. From within this directory, this model can be compiled using

```
ncvlog -ams vco.vams
```

or in case the working library is not preset or it should be compiled into a library other than the default working library

```
ncvlog -ams -work worklib vco.vams
```

The command line argument "*-ams*" is necessary to compile Verilog-AMS models. Compilation without this argument compiles Verilog-HDL models only. When compiling several source files the order in which they are compiled is arbitrary. If there are syntax errors, *ncvlog* will print error messages that direct you to the point in the file where the error was detected. For a quick command line reference you can use

```
ncvlog -help
```

An example PLL-design is shown in Listing 4; it references modules defined in Listings 5-8. Every module is written into a separate file in the working library. The design is compiled with the compilation script shown in Listing 9. The phase-frequency detector module *(pfd)* and the frequency divider module *(fd)* are compiled without the **-ams** switch. These modules are purely digital and written in Verilog-HDL.

LISTING 4 *The top-level module of the PLL example (from file plltop.vams).*

```
`include "disciplines.vams"
`timescale 10ps / 1ps

module plltop ();
    electrical gnd;
    ground gnd;
    reg ref, reset;

    initial begin
        ref = 0;
        reset = 1;
        #100 reset = 0;
    end

    always #3333 ref = ~ref; // 15MHz

    pfd PFD (.reset(reset), .qinc(up), .active(fb), .ref(ref), .qdec(dwn));
    cp #(.cur(1m)) CP (.nout(gnd), .dec(dwn), .inc(up), .pout(err));
    capacitor #(.c(30n)) C (err, err2);
    resistor #(.r(200)) R (err2, gnd);
    vco #(.f0(1.5E9), .kvco(50.0E6), .rin(100k)) VCO (.ps(err), .ns(gnd), .out(out));
    fd FD (.reset(reset), .out(fb), .clk(out));
endmodule
```

4.1.3 Elaboration

During the elaboration step all models that are used in the design are linked together. The hierarchy of the design is built. Before starting the elaborator, it is necessary to compile all models that will be used. Then the elaborator can be invoked with

```
ncelab plltop -discipline logic -timescale 10ps/1ps
```

where *plltop* is the name of the top level module in the design. The elaborator finds the names of the instantiated modules within the compiled *plltop* module. It runs the discipline resolution and the connect module insertion algorithms. The **-discipline** command line option is used to define the default discipline. Defining it as *logic* allows the use of Verilog-HDL modules, like the *pfd* and *fd* modules, to be used without adding a discipline declaration for the logic signals to the module source code. The **-timescale** attribute assigns the time unit and time precision information to modules that do not have *'timescale* directives.

LISTING 5 *The phase-frequency detector (from file pfd.v).*

```
`timescale 10ps / 1ps

module pfd (qinc, qdec, active, ref, reset);
    output qinc, qdec;
    input reset, active, ref;
    wire fv_rst, fr_rst;
    reg q0, q1;

    assign fr_rst = reset | (q0 & q1);
    assign fv_rst = reset | (q0 & q1);

    always @(posedge active or posedge fv_rst) begin
        if (fv_rst) q0 <= 0; else q0 <= 1;
    end

    always @(posedge ref or posedge fr_rst) begin
        if (fr_rst) q1 <= 0; else q1 <= 1;
    end

    assign qinc = q1;
    assign qdec = q0;
endmodule
```

Invoking the elaborator with the above command line options applies the default basic mixed signal discipline resolution mechanism. If the detailed discipline resolution mechanism is desired, the command line switch **-dresolution** must be added.

More information about the steps taken during the elaboration process is displayed if the **-messages** command line switch is added to the command line.

4.1.4 Simulation

The simulation can be run in batch mode, in TCL command mode or using the graphical user interface (GUI). In the TCL and GUI modes the user can interactively communicate with the simulator. To simulate the PLL example design in GUI mode the simulator is invoked with

```
ncsim plltop -analogcontrol plltop.scs -gui
```

To start the simulator in TCL command mode, the **-tcl** command line option is used instead of **-gui**. The simulator starts in batch mode if neither of these two command line options is present. When running in batch mode it is possible to control the simulation with a TCL control script that might contain, among other instructions, signal probe commands for later viewing of the result waveforms. The TCL control script can be referenced using the **-input** command line option:

LISTING 6 *The charge pump (from file cp.vams).*

```
`include "disciplines.vams"
`timescale 10ps / 1ps

module cp (pout, nout, inc, dec);
    parameter real cur = 1m;        // output current (A)
    input inc, dec;
    electrical pout, nout;
    real out;

    analog begin
        @(initial_step) out = 0.0;

        if (dec && !inc)
            out = –cur;
        else if (!dec && inc)
            out = cur;
        else out = 0;

        I(pout, nout) <+ –transition(out, 0.0, 10n, 10n);
    end
endmodule
```

```
ncsim plltop -analogcontrol plltop.scs -input control.tcl
```

The file *plltop.scs* is the control file for the analog simulation engine. In the simplest case this file would contain a transient analysis statement

```
transient tran stop=1ms
```

which defines the stop time for this simulation run.

Starting the PLL example in GUI mode will invoke the *SimVision* simulation control and debug environment. In the *Design Browser* window select the signals you want to probe, then hit the *"Send To Waveform"* icon at the right side of the icon bar. The SimVision waveform window opens and shows the still empty waveforms for the selected signals. Now, in the waveform window hit the *"Run Simulation"* icon (white arrow on black). The simulation will run up to the stop time defined in the transient statement within the analog control file. But at any time the user may pause the simulation and continue or forward it step by step for debug purposes. The Cadence product documentation provides details about the features used to debug a model.

4.1.5 Automatic Interface Component Insertion

The PLL example presented above does not require interface component insertion because the disciplines of the interconnecting wires match. Now lets assume we want

LISTING 7 *The frequency divider (from file fd.v).*

```
`timescale 10ps / 1ps

module fd (out, clk, reset);
    input clk, reset;
    output out;
    wire out;
    reg q;
    integer i;

    always @(negedge reset) begin
        i = 0;
        q = 0;
    end

    always @(posedge clk) begin
        if (~reset) begin
            i = i + 1;
            if (i == 63) begin
                q = ~q;
                i = 0;
            end
        end
    end

    assign out = q & ~reset;
endmodule
```

to use an analog sine wave signal as the PLL reference clock instead of the digital oscillating signal *ref*. Listing 10 shows the changed top level netlist. A sine wave voltage source (*V1*) generates the analog signal *ref*. This analog signal is connected to the *ref* input of the phase-frequency detector (*PFD*), which is modeled completely in Verilog-HDL and so has a digital input. Obviously a mixed-signal interface component is necessary in this case.

A library with connect modules and corresponding connect rule definitions is provided with the AMS Designer simulator. The user chooses one of the connect rule definitions and references them when invoking the elaborator. With *mixedsignal_2* as the chosen connect rule definition, the elaboration command line changes to

```
ncelab plltop mixedsignal_2 -discipline logic -timescale 10ps/1ps
```

The additional elaborator command line option **-iereport** displays information about the inserted connect modules. The Verilog-AMS language also provides users with the ability to create their own connect modules. If those should be used instead of the

LISTING 8 *The VCO (from file vco.vams).*

```
`include "disciplines.vams"
`timescale 1ns / 1ps

module vco (out, ps, ns);
    parameter real f0 = 100k;                  // center frequency (Hz)
    parameter real kvco = 10k;                 // gain (Hz/V)
    parameter real rin= 100k from (0:inf);     // input resistance (Ohms)
    output out;
    electrical ps, ns;
    reg out;
    logic out;
    real vin;

    initial out = 0;

    always begin
        vin = V(ps, ns);
        #(0.5e9 / (f0 + kvco * vin))
        out = ~out;
    end

    analog I(ps, ns) <+ V(ps,ns)/rin;
endmodule
```

LISTING 9 *Compilation script (from file compile.sh).*

```
ncvlog -ams cp.vams
ncvlog -ams vco.vams
ncvlog pfd.v
ncvlog fd.v
ncvlog -ams plltop.vams
```

provided connect modules, they must be compiled into a library. This library has to be defined in *cds.lib*

```
define my_connectlib ./my_connectlib
```

and the library directory, in this case *./my_connectlib* must exist. The connect modules as well as the connect rules block are compiled with

```
ncvlog -ams -work my_connectlib my_connectmodules.vams
ncvlog -ams -work my_connectlib my_rules.vams
```

The name of the connect rules must be specified in the elaboration command line

LISTING 10 *PLL top level netlist with an analog reference clock.*

```
`include "disciplines.vams"
`timescale 10ps / 1ps

module plltop ();
    electrical gnd;
    ground gnd;
    reg reset;

    initial begin
        reset = 1;
        #100
        reset = 0;
    end

    vsource #(.type("sine"), .ampl(2.5), .dc(2.5), .freq(15M)) Vin (ref, gnd);
    pfd PFD (.reset(reset), .qinc(up), .active(fb), .ref(ref), .qdec(dwn));
    cp #(.cur(1m)) CP (.nout(gnd), .dec(dwn), .inc(up), .pout(err));
    capacitor #(.c(30n)) C (err, err2);
    resistor #(.r(200)) R (err2, gnd);
    vco #(.f0(1.5E9), .kvco(50.0E6), .rin(100k)) VCO (.ps(err), .ns(gnd), .out(out));
    fd FD (.reset(reset), .out(fb), .clk(out));
endmodule
```

```
ncelab plltop my_rules -discipline logic -timescale 10ps/1ps
```

4.1.6 Design Configuration

Design configuration enables the easy use of modules with different abstraction levels. Besides the behavioral version of the VCO model given above, there might be another one, also named *vco*, that includes a transistor level description. How does the tool select which one to use for simulation?

There are two different ways for configuring the design. When the AMS Designer environment is used, a configuration file for the design is generated and fed into the simulator. When running the simulator from the command line this file could be created manually. However there is a second possibility that is easier to use when working on the command line: binding. Using this method the elaborator identifies the right module implementation based on the name of the library the module is compiled into and the name of a view. This means the two differently described VCOs could be in different libraries, or they could be even in the same library but with different view names. Design configurations are applied during the elaboration step. However, during compilation of the modules the library and the view have to be specified. The compilation with

ncvlog -ams -work libA **-view** behavioral vco.vams

puts the compiled module into the library *libA* and assigns the view name *behavioral* to it. Another VCO with the same module name could be compiled into the same library under a different view name, for example *schematic*. The elaboration command

ncelab plltop **-binding** libA.vco:behavioral

specifies exactly which module implementation to use. If there are several plltop modules available, library and view can be specified for the top level module:

ncelab libA.plltop:detailed

In this case the view was named *detailed*. When there are several top level modules with the same name, the desired one needs to be specified on the simulator command line as well:

ncsim libA.plltop:detailed **-analogcontrol** plltop.scs **-gui**

4.2 Referencing SPICE

SPICE as well as Spectre primitives and subcircuits can be instantiated within Verilog-AMS modules as described earlier in this appendix. Simulator built-in primitives are found automatically during the design elaboration. However, the location of SPICE or Spectre model statements and subcircuit definitions must be specified. This can be done in the *hdl.var* file using the *MODELPATH* declaration:

define MODELPATH <unix_path_to_netlist>

or on the elaborator command line with the command line attribute

-modelpath <unix_path_to_netlist>

4.3 Referencing VHDL-AMS

The AMS Designer simulator supports VHDL-AMS as an additional mixed signal modeling language. VHDL-AMS models can be instantiated in Verilog-AMS netlists like any Verilog-AMS module. Before the design elaboration step VHDL-AMS models (*entity* and *architecture*) must be compiled. For a model in a file named *vco.vhd* one would use:

ncvhdl -ams vco.vhd

Both Verilog-AMS and VHDL-AMS can be compiled into the same library. For discipline resolution and automatic interface component insertion at the mixed language boundary, the *electrical* Verilog-AMS discipline is compatible with the VHDL-AMS *electrical* nature, and the *logic* discipline is compatible with *std_logic* in VHDL-AMS. Automatic interface component insertion at the language boundary follows Verilog-AMS rules.

References

[1] J. Bhasker. A Verilog HDL Primer (2nd edition). Star Galaxy Publishing, March 1999.

[2] Cadence Design Systems. AMS Designer, *www.cadence.com.*

[3] N. Chandra, and G. W. Roberts. Top-down analog design methodology using Matlab and Simulink. *Proceedings of the 2001 IEEE International Symposium on Circuits and Systems* (ISCAS '01), vol. 5, 2001, pp. 319-322.

[4] E. Chou and B. Sheu. Nanometer mixed-signal system-on-a-chip design. *IEEE Circuits and Devices Magazine*, vol. 18, no. 4, July 2002, pp. 7-17.

[5] M. D. Ciletti. *Advanced Digital Design with Verilog HDL.* Pearson Education, August 2002.

[6] CoWare, Inc. Signal Processing Worksystem, *www.coware.com.*

[7] C. A. Desoer, E. S. Kuh. Basic Circuit Theory. McGraw Hill, 1969.

[8] D. Fitzpatrick and I. Miller. *Analog Behavioral Modeling with the Verilog-A Language*. Kluwer Academic Publishers, 1998.

[9] C. Force, T. Austin. Testing the design: the evolution of test simulation. *International Test Conference*, Washington 1998.

[10] C. Force. Integrating design and test using new tools and techniques. *Integrated System Design*, February 1999.

[11] J. E. Franca. Integrated circuit teaching through top-down design. *IEEE Transactions on Education*, vol. 37, no. 4, Nov. 1994, pp. 351-357.

[12] G. G. E. Gielen. Modeling and analysis techniques for system-level architectural design of telecom front-ends. *IEEE Transactions on Microwave Theory and Techniques*, vol. 50, no. 1, part 2, Jan. 2002, pp. 360-368.

[13] G. G. E. Gielen. System-level design tools for RF communication ICs. URSI International Symposium on Signals, Systems, and Electronics (ISSSE '98), 1998, pp. 422-426.

[14] J. Holmes, F. James, and I. Getreu. Mixed-Signal Modeling for ICs. Integrated System Design Magazine, June 1997.

[15] S. A. Huss. *Model Engineering in Mixed-Signal Circuit Design: A Guide to Generating Accurate Behavioral Models in VHDL-AMS*. Kluwer Academic Publishers, January 2002.

[16] *IEEE Standard Hardware Description Language Based on the Verilog Hardware Description Language*. IEEE Standard, 1364-1995.

[17] K. Kundert. Automatic model compilation, an idea whose time has come. *www.designers-guide.com/Opinion*.

[18] K. Kundert, H. Chang, D. Jefferies, G. Lamant, E. Malavasi, and F. Sendig. Design of mixed-signal systems-on-a-chip. *IEEE Transactions on Computer-Aided Design of Integrated Circuits and Systems*, vol. 19, no. 12, Dec. 2000, pp. 1561-1571.

[19] K. S. Kundert. *The Designer's Guide to SPICE and Spectre*. Kluwer Academic Publishers, 1995.

[20] K. Kundert. *Modeling Varactors*. *www.designers-guide.com*.

[21] Mathworks. Matlab and Simulink, *www.mathworks.com*.

[22] T. Murayama, Y. Gendai. A top-down mixed-signal design methodology using a mixed-signal simulator and analog HDL. *Proceedings EURO-DAC '96, The European Design Automation Conference and Exhibition*, 1996, pp. 59-64.

[23] S. Palnitakar. *Verilog HDL* (2nd edition). Prentice Hall, March 2003.

[24] G. Peterson, P. Ashenden, D. Teegarden. *The System Designer's Guide to VHDL-AMS*. Science & Technology Books, September 2002.

[25] A. V. Oppenheim, A. S. Willsky, H. Nawad. *Signals and Systems* (2nd edition). Pearson Education, 1996.

[26] Teradyne. SpectreVX and SaberVX virtual test environments, *www.teradyne.com*.

[27] D. E. Thomas, P. R. Moorby. *The Verilog Hardware Description Language* (5th edition). Kluwer Academic Publishers, May 2002.

[28] *Verilog-AMS Language Reference Manual: Analog and Mixed-Signal Extensions to Verilog-HDL, version 2.1*. Accellera, 2003. Available from *www.accellera.com*. An abridged version may also be found at *www.verilog-ams.com*.

[29] Verilog-AMS library, *www.verilog-ams.com*.

[30] R. E. Ziemer, W. H. Tranter, D. R. Fannin. *Signals and Systems: Continuous and Discrete* (4th Edition). Prentice Hall, 1998.

Index

Symbols

A

W

X

Z